In the Beginning
Was the Worm

Also by Andrew Brown

Watching the Detectives

The Darwin Wars

In the Beginning Was the Worm

Finding the Secrets of Life
in a Tiny Hermaphrodite

ANDREW BROWN

SIMON & SCHUSTER

London . New York . Sydney . Tokyo . Singapore . Toronto . Dublin

A VIACOM COMPANY

First published in Great Britain by
Simon & Schuster UK Ltd, 2003
A Viacom company

1 3 5 7 9 10 8 6 4 2

Simon & Schuster UK Ltd
Africa House
64–78 Kingsway
London WC2B 6AH

www.simonsays.co.uk

Simon & Schuster Australia
Sydney

A CIP catalogue record for this book
is available from the British Library

ISBN 0–7432–0716–5

Typeset by M Rules
Printed and bound in Great Britain by
The Bath Press, Bath

Contents

Acknowledgements

I thank all the people who talked to me, especially Sydney Brenner, Alan Coulson, Bob Horvitz, Donna Albertson, John Sulston, John White, Leon Avery, Bob Edgar, Judith Kimble, Phil Anderson, Bob Waterston, Nichol Thomson, Eileen Southgate, Jonathan Hodgkin, Marty Chalfie, David Hirsh, Richard Durbin, Muriel Wigby and Barbara Meyer. Joan Green, the librarian at the Sanger Centre, was extremely kind and helpful. I am conscious that Andy Fire should have played a more prominent part in the book, and he would have done, had I been able to arrange a trip to see him.

I am grateful also to Caroline Brown, who typed acres of transcripts and read carefully every word, to my great benefit, and Rosamond and Felix Brown who were patient while all this went on. Joze Stare and his family provided tranquillity and wonderful views to write in front of. Helen Simpson was, and remains, the nitpicker from hell.

Introduction

Once upon a time John Sulston was talking to two teenage girls on a train. He was a bearded hippie type, but friendly and serious: a safe man to talk to. Still, when he told them about his work, they giggled ferociously, because he explained that he had spent ten years or so dissecting in intimate detail tiny, transparent, hermaphrodite nematode worms. He wasn't offended by their laughter. He had often argued with his friends that it was absurd for scientists to be paid more than dustmen, not because dustmen were more useful, but because scientists had so much more fun.

At a press conference about fifteen years later, the same man, trimmed all round to look respectable, but still serious and friendly enough to be a really effective revolutionary, sat alongside Tony Blair to hear the prime minister describe his work as 'a revolution in medical science whose impact can far surpass the discovery of antibiotics' and 'the first great technological triumph of the twenty-first century'.

This was modest compared to the estimate of Sulston's immediate boss, Mike Dexter, who reckoned that 'This is the outstanding achievement not only of our lifetime but in terms of human history . . . this code is the essence of mankind.'

Blair and Mike Dexter were hyping the human genome project, not the worm on which Sulston had spent most of his life working. But the worm's genome had been sequenced first; in 1998, it became the first multi-cellular organism to have all its DNA listed and read. Sulston, his friend Bob Horvitz and his former boss Sydney Brenner, who invented and drove the worm project, shared the Nobel prize for physiology and medicine in 2002 for their worm work, not for the human genome. The human genome project grew directly from the work that had been done on worms. It would have been done eventually without the worm; but it would have been done much more slowly, perhaps less thoroughly; and certainly less publicly, and so far less usefully. Very probably, without Sulston the genome sequence would have been locked up and almost completely useless in the vaults of some American biotech company.

So you can't understand the story of the human genome without looking at the story of the worm. Neither can you really understand the culture and the attitudes among scientists that made the human genome project look worthwhile without looking at the cultures that shaped them; and the worm formed an important part of that. Its history spans almost the whole of molecular biology, which has gone in fifty years from being a field without textbooks, or even a name, to the gigantic multi-billion-dollar industries of today.

Anyone listening to the hype about the human genome could be pardoned for wondering, Where is the benefit? For all the thousands of millions of pounds spent on the enterprise, not a single cure for a single disease has been found directly from the sequence. Even the language of news reports has changed so that they now talk about 'a gene that contributes to' some condition, rather than 'the gene for'. The stock price of Celera, the company that produced a private version of the human genome sequence, fell from $240 to $10 in less than three years; and we are told that the effort (and money) needed to decode the genome will be dwarfed by the effort and money needed to examine the proteins that the genome specifies. Only when they are all understood will all the promised benefits to humanity appear.

The worm story helps to understand why this has happened. No one stood to make a fortune from the study of nematode worms. No one cared or cares for the soul of a nematode (though there is a rumour that the cell containing it was found and discarded by researchers in 1978). Their motives for reading the worm's genome were purely scientific. Only by understanding why and how it was done to the worm can one hope to understand the real reasons for sequencing the human genome, and the real benefits that might accrue from it.

But the importance of the worm goes deeper. This tiny scrap of brainless voracity is far more complicated than it seems. Learning about it has told us a great deal about the fundamental mechanisms of our own bodies. Anything alive can reproduce and eat, and the overwhelming majority of living things on this earth are bacteria. But the life that we notice – which means anything large enough to see – can do more than reproduce and eat or even move. It can grow. When its cells divide, they don't split off into new organisms, but stay linked with the old ones, to make complex, cooperating wholes.

This is so complex and so difficult that most of the genes of any multi-cellular creature are occupied with it: that's why we share about 50 per cent of our genes with a banana.

A banana and a human being – or a worm – are very differ-ent. But they are all recognisable, coherent, organised wholes, and this similarity turns out to be almost the most important thing about us. It is certainly the foundation on which all the subsequent differences are built.

The worm, *Caenorhabditis elegans*, is about as simple as an animal can be. It has only 959 cells that are not eggs or sperm, and all these are known, and all grow in the same way. They are identical from worm to worm. In many ways, the worm is nature's test tube: a transparent organism which might have been created to do experiments in, not least because it has no brain and can't be imagined suffering. But none of this explains why one should want to do all these experiments, and why people so burned to understand.

It is here that the worm shows us most about our own species. Modern science has taught us an enormous amount about the world around us, and extraordinarily little about the worlds within us. There is certainly no scientific expla-nation for scientific passion, but the history of the worm is full of the devouring lust for knowledge. The people who did it were not really interested in money, or fame outside a limited circle. They were not saints. They were ambitious and competitive, and life was hard for those who failed. But their ambition and competitiveness and their sometimes jealous love were all directed at altruistic ends. They wanted to understand the world. They wanted measurable, solid truths about it. Their Nobel prize was recognition, not the reward they had worked for.

This kind of passion looks extraordinary because it is both ruthless and unselfish. We think that if a businessman works twelve- or eighteen-hour days, neglecting his family, and careless of the world's opinion, this makes sense because he will be rich at the end. But the worm scientists worked like that with no expectation of becoming rich.

Indeed, very few of them did so, and some of the most important ones, among them John Sulston, deliberately turned their backs on great wealth in order to be able to share their discoveries with the world. I don't think that studying our DNA will tell us nearly as much about what it means to be human as understanding that kind of passion would.

This purity and focus of vision runs right through the story of the worm, despite the increasing elaboration and expense of the science involved. In some ways, the tiny worm stretches across the whole history of molecular biology, from the original experiments on the smallest possible living things, to the final ambition to have sequenced the DNA of everything on the planet – because only then can we properly understand it. There may be a huge change in outward circumstances between the original experiments, conducted in scrounged lab space all round Cambridge, and the global reach of the business now, conducted in hi-tech purpose-built campuses all round the world. But there is a single common thread of a focused ambition running through, and a determination to understand how it is that certain chemicals assemble themselves into stable patterns, and how these patterns combine in the exquisitely complicated and patterned chemical dance of life.

It is an extraordinary story, and all the more extraordinary because all this passion was driven and directed by one man's vision.

1

Sydney Brenner

Morris Brenner was illiterate all his life, though he spoke Yiddish, Russian, Afrikaans and two African languages as well as English. His son, Sydney, learned to read at the age of four, from poverty as much as anything: he was being looked after by a widow who lived in one room, and who spread her table with old newspapers because she could not afford a cloth, and Sydney was simply taught by looking at the patterns the ink made on the paper. Morris was a cobbler, part of the vast Jewish diaspora from tsarist Russia, who ended up outside Johannesburg when he emigrated in 1910 because the fare there was cheaper than to America. In South Africa he married Lena Blacher, a Latvian Jewish émigrée. Their second son, Sydney, was born in 1927.

Sydney Brenner was taken into a kindergarten as a charity child after he had learned to read; he went up to Witwatersrand University to study medicine at the age of fourteen, a recognised prodigy, three years younger than anyone else in the class. He was so enchanted by the first biology textbook he borrowed from the library that he claimed to have lost it and paid the fine rather than return it. He still has little time for the idea that you can only learn if formally taught. 'I just never had that experience. So I cultivated, probably out of necessity, but certainly combined with inclination, the idea that knowledge is out there, it's available. If you can't buy the book, you can always go to the public library. And steal the thing, if necessary!'*

He had supported himself through university partly with a scholarship but also as a professional mourner: his job was to be the tenth man required at prayers for the dead. The only disadvantage of this precocity appeared when he was ready to qualify as a doctor and discovered that, at twenty, he'd be too young legally to practise. To fill in the time, he took a year's courses towards a degree in anatomy and physiology, which interested him rather more than practical doctoring. The only part of his medical studies at which he excelled was obstetrics and gynaecology, after a stint in a black hospital in Durban, staying with three other students in a hotel which doubled as a brothel. 'There was absolutely nothing to do except learn how to deliver babies; since I had to do it, I managed to do it extremely well.'

The worm itself looks rather less interesting than the scientist. In fact, Brenner's friend the embryologist Lewis Wolpert once called it the most boring animal imaginable. It is about half a millimetre long, transparent, and lives its natural life in soil or, for a treat, in compost and dung heaps. One of the minor mysteries of the story is how it came about that anyone ever noticed the worm's existence, though it has been known to science since 1899. In a laboratory the worms are kept in petri dishes, transparent plastic ramekins with a lid; and when you look at one of these, all you can see at first is a clear jelly in the middle, which contains the bacteria they eat. Then, if you peer closely, it looks as if

* Sydney Brenner and Lewis Wolpert, *A Life in Science*, p. 3

there is a minute sprinkling of hair on the jelly; that's all the naked eye discerns. But placed under a microscope and illuminated from below, the worm is transformed. It's beautiful. There is an elegant double taper to the creature, and it moves with a sinuous determination, leaving curving tracks through the agar as a microscopic skier might. That's why its name is *Caenorhabditis elegans*. Magnified and illuminated, the worm also develops subtleties of transparency. I first saw it through a blue filter, where it is simply glorious. The outer skin is a rather aquatic blue, and inside the organs are visible as darker blue shadings or lines. A pregnant worm fills slowly with eggs like glowing blue marbles. If you want to understand why really intelligent people with a love of life and of society were happy to obsess for months about worms, watching them for eighteen-hour days and then waking in the night with fresh ideas about them, you have to see that the worm is not just a transparent lens through which the rest of biology may be studied. It is beautiful in itself.

Brenner is not beautiful in repose. But the force and intelligence of his features are astonishing. Even as an old man, almost bald, with eyebrows that burst out like the roots of a small forest, he seems to fill a room with vigour. His eyes are a curious bluey-green (a colour I have only otherwise seen in trout streams in winter in the Slovenian Alps). When he starts to talk you are swept along in the icy, buffetting current of ideas, shocked and exhilarated to the point of exhaustion – and still he goes on talking. Profundities, puns, anecdotes and opinions all rush and jumble together. 'The only time I was able to talk science uninterruptedly with Sydney came when he was in hospital after a bad motorcycle smash,'* said one of his collaborators

* Phil Anderson, interview with the author. (Except where indicated otherwise, all quotes are from interviews.)

years later. Otherwise, he will talk about almost anything. 'Was Grandfather born stupid, or did he have to work at it?' one grandchild asked his grandmother, bewildered by the rush of puns.

Good scientists must have a Promethean faith in their ability to bring back knowledge from the world around them; but in the Cambridge milieu where the worm research started, people would rather fart than boast out loud. Overt self-advertisement would be a grotesque solecism in that atmosphere of unselfconscious, understated elitism. Indeed, some of the best of the worm people go to extraordinary lengths of self-effacement. When I asked one man, innocently, what had led him into biology, he replied, 'Well, I came from a medical family.' This turned out to mean 'My father won a Nobel prize.' Fred Sanger, the only man ever to have won two Nobel prizes in the same discipline, is one of the most mild and unassuming men I have ever met. John Sulston, whose efforts on the worm and then on the human genome really have changed the course of history, didn't want to be profiled by a newspaper when his book came out, for fear that people would think him showy.

Brenner is quite without such diffidence. In 1963, when he decided to spend the next decades of his life studying a minute, transparent hermaphroditic worm, he was thirty-six and already one of the heroes of molecular biology for his part in cracking the genetic code. That he had to wait another thirty-nine years for his Nobel prize was, thought most people who knew him, one of the mysteries of modern science.

After finishing his studies in South Africa, he had come to Oxford on a research fellowship in 1952. In the wintry spring of 1953, with a couple of friends, he crammed into a small unheated car and drove across to Cambridge to inspect the famous model that Jim Watson and Francis

Crick had built of DNA, and from then on he just threw himself into the problems that it raised. His brilliance, his capacity for work, his confidence and ambition, propelled him to the heart of the group that worked out the way in which a long string of DNA, consisting of only four elements endlessly repeated in slightly varying orders, was translated into twenty amino acids, which are in turn strung together to make proteins, which in their turn combine to make worms, and us, and all other living things.

Everyone knows of Crick and Watson's discovery of the structure of DNA. But that didn't crack the code: it merely showed where the code is written; and discovering where a coded message has been hidden is quite different from reading it. Crick and Watson had shown that the information in genes must somehow be contained in the endlessly varied arrangement of four different chemical letters along the inside of a double spiral of DNA. Working out how the information was actually encoded in these sequences and then read out from them took another ten years of concentrated effort around the world.

Before DNA, Brenner said later, biology had studied three-dimensional beings; but once it was grasped that the information needed to build any living thing is contained inside the double spiral of DNA, biology could be reduced to only one dimension: the almost endless string of chemical letters along the DNA, which would be used to disentangle every other problem in biology. 'Once we had this absolutely clear-cut conception then, in this very small evangelical sect, we realised that everyone else was simply talking nonsense.'*

In 1958, Crick and Brenner coined the term 'molecular genetics' to describe what they were doing. There was a

* Brenner and Wolpert, *A Life in Science*, p. 12

huge ambition contained in those two words, whose true
size only gradually become apparent over the next five
decades. They would end up taking apart the whole of biology to reassemble it on a basis of an understanding of what
goes on in cells at the smallest and most detailed level.

But by 1966 the foundations were firm. A series of
experiments of extraordinary simplicity and penetration,
conducted using the smallest and most primitive beings
imaginable – viruses which prey only on bacteria – had
solved the code in principle and then painstaking biochemistry solved the rest in practice. Given a sequence of
bacterial DNA, it was possible to predict exactly which proteins might be made from it. This was all the more
impressive because it was, in a sense, a purely theoretical
discovery.* Huge deductions about the fundamental structure of all living things were made by watching the
discolourations of splodges of bacteria growing in a grid of
petri dishes. Just as the fundamentals of genetics had been
worked out long before anyone knew what genes were
actually made of, so most of the ways genes work together
were discovered long before anyone had sequenced a
length of DNA to discover exactly which bases lay in what
order.

But however deep the intellectual reach, what held this
new molecular biology together at its base was a very simple
idea. All the properties, and ultimately all the behaviour, of
all living things could be explained as a sort of immensely

---

* Indeed, with growing practical knowledge about the DNA of
creatures more complicated than bacteria, it got to some extent
undiscovered again. The problems of introns, and alternative
splicings, mean that you have to know not only what the DNA says,
but whether the cell is listening when the DNA speaks. But that will
all emerge in a later chapter.

complicated Meccano. Living things are made of trillions of unliving molecules; and in the last analysis all life consists of the ways in which these molecules move around each other, sometimes fitting together, sometimes separating or migrating; and all that complex dance is in turn determined by a few very simple properties like their electrical charges, which can be deduced from the first principles of quantum mechanics. If you know the shape of the molecules on a really fine scale – so small that they are literally invisible even to the most powerful microscopes, and have to be investigated using radiation finer than light – all the other qualities of living things should be explicable in terms of how these shapes fit together or move apart.

You can watch this shift of understanding very clearly in the ways that we think about breathing. The association of life and breath goes back about as far as any idea can be traced in history. In ancient Greek the two words were the same, and I don't think there can be a culture anywhere in the world where people don't understand that when you stop breathing you are dead. People have known that for as long as they have been people, and probably for a few hundred thousand years before then. But for almost all that time the understanding was that breathing was something almost independent of the body. Your breath could leave you, and pursue an independent existence elsewhere. Nor was breath put together with the other, equally widespread, understanding that without enough blood you will die. The fact that blood circulates endlessly round the body from the heart was not discovered until 1613; the existence of oxygen was not discovered till 1785; the two thoughts were not put together until the nineteenth century. But by 1940 people understood that what makes blood red is the protein haemoglobin, which somehow carries oxygen round the body and that oxygen is necessary to fuel the chemical reactions which are what distinguish living bodies from dead ones.

But in the 1940s Max Perutz, an Austrian émigré in Britain, began to wonder why haemoglobin transports oxygen: what is it about the protein which enables it move oxygen round the body and deliver it to the places where it is needed? The answer to this last question turns out to reside at the sub-microscopic level of the individual molecule. You can see a single red blood cell clearly enough under a microscope but the haemoglobin molecule is about six orders of magnitude smaller: there are around 280,000,000 of them in every blood cell, and of course there is an unthinkably larger number of blood cells in every human being: 25 trillion of them, give or take a few thousand billion.

By bombarding haemoglobin crystals with X-rays you could hope that some would be knocked out of their course by the molecules within; by mapping carefully where these diverted X-rays ended up, you could get a pattern from which it was possible, by the application of complex mathematics – months and months of calculation in the days before computers – to work out where the individual molecules should be. It is an extraordinarily indirect procedure. A geneticist friend of mine once compared it to unravelling a sweater, photographing the resulting ball of wool, and then trying to reconstruct the original knitting pattern from these photographs.*

X-ray crystallography was used to work out the structure of DNA, but haemoglobin is a far more complex molecule. In fact, it took Max Perutz thirty years of monotonous, patient labour to understand the molecule completely. At the end of that time, however, he had teased out the whole monstrous,

* Jackie Leach Scully, Swarthmore lectures 2002.

tangled structure. It looks something like a very battered armchair modelled from four bits of macaroni; and he had seen exactly how the structure must shift very slightly to trap two molecules of oxygen when the haemoglobin molecule is in the lungs, and shift again to squeeze them out in response to the different pressure and acidity of the small blood vessels that nourish our muscles. The most important word in the previous sentence is 'must'. Perutz could show how haemoglobin must do its work because it follows the same physical and chemical laws as explain how stars must manufacture inside themselves all the elements we and our world are made of. And this is so extraordinary it's hard to look at straight.

By 1965, when the worm project started, scientists had managed to integrate the way living things are put together into their understanding of everything else in the universe. We live in a world where genes seem familiar and well understood – no odder than the fact that Australians don't fall off. But the roots of molecular biology are, in fact, at least as odd as the idea that our stable world isn't flat at all but is actually whirling through space at giddy speeds all the time.

The chains of protein, like macaroni, that make up a haemoglobin molecule, are made of smaller links called amino acids. There are twenty different kinds of amino acid which are used to build everything that lives on earth: as you add different links to each chain, so it folds up into a different tangle, and these infinitesimal tangles of protein are what we're all built from. What specifies the order of amino acids and so, ultimately, the particular way a protein tangles into its final form is the pattern of chemical letters along a strand of DNA. A single change in one of these letters can change the amino acid it specifies: when that amino acid is built into a new haemoglobin molecule, the shape is subtly different. It no longer changes shape as it should when it reaches the capillaries: instead, these defective haemoglobin molecules clump together in their millions, distorting the red blood cells into a lumpy crescent shape.

Bad copies of the haemoglobin gene that have this effect are surprisingly common. If your original fertilised egg had one such copy from a parent, all the trillions of blood cells in your body will have it, too, and may make this duff haemoglobin; but it won't matter because the other half will make the sort that works properly. Half your eggs will have this dud copy, too: but if you find a man in the same condition, so half his sperm have this subtly wrong version, or allele, of the haemoglobin gene, there may be dreadful trouble. For if the wrong sperm and the wrong egg get together, the child who grows from the fertilised egg will have none of the right sort of haemoglobin and will slowly die of a disease known as sickle-cell anaemia. Actually, there are many of these diseases of haemoglobin, all caused by changes in amino acids which disrupt the normal way in which the protein traps or releases an oxygen molecule. You would expect them to be eliminated by natural selection, but it appears that the distorted haemoglobin molecules are not worthless. They are abnormally inhospitable to the malaria parasite. So, if you live in a region where malaria is endemic, having half your haemoglobin like that may keep you alive, even if it condemns a quarter of your children. In some conditions of infant mortality, that's a pretty good bet.

This may seem a long way from worms. But it is a classic example of how a molecular understanding can fit into the chains of causes and consequences of the world around us, and make the workings of our bodies clear enough for us perhaps actually to be able to do something about them.

By 1963, it seemed obvious to the fifty or a hundred people who had understood what had happened that, once molecular biologists had understood the meaning of the genetic code, and the way that cells read information off DNA, they understood, in principle, everything and that 'Everyone else was simply talking nonsense.'* Forty years before the announcement of the triumph of the human genome project, it seemed to the very smartest people in the field that everything from then on was a working-out of details. It's a good general rule that the educated newspaper reading public is about thirty years behind the understanding of the people doing cutting-edge research. So if you try to remember the euphoria that greeted the end of the human genome project – the announcement by President Clinton that we had finally read the book of life and all that sort of thing – you have some sense of how complete the molecular biologists felt that their triumph had been in the early Sixties. They thought they had made the rest of science completely redundant. 'Anyone who would hire an ecologist is out of his mind,' said Jim Watson at a Harvard faculty meeting in the early Sixties, and he meant it.

So most of them moved on to colonise new worlds. It was in the nature of molecular biology that it had been mostly done by people who weren't biologists by training, because it was a field which had to be invented from scratch. A training in conventional biology was less use in working out the structure of DNA than a background in hard physics; and the 'phage' – the viruses which infect bacteria, and which were used to work out the genetic code – are so simple that it's reasonable to ask whether they are alive at all. Many of the first molecular biologists had, like Crick himself, been physicists. Watson had been an ornithologist.

* Brenner and Wolpert, *A Life in Science*, p. 12

Brenner had got his first degree as a medical doctor. Now they spread back out across the rest of biology, confident that it had nothing to teach them (the Harvard evolutionary biologist Ed Wilson, who first met Watson in this period, thought him then 'the most unpleasant human being I had ever met', who 'radiated contempt in all directions at department meetings'*).

After his time at Oxford, Brenner had eventually been forced to return to South Africa by the terms of his scholarship; but in 1957 Crick got him a place at the Laboratory of Molecular Biology (LMB) in Cambridge, which was probably at that moment the most advanced place in the world for studying the field that Crick and Brenner had named molecular genetics. Brenner returned with a wife and two children, on a one-year contract; but there was never any question of sending him back. Within five years, he and Crick believed that they had skimmed all the cream from the subject and they were looking for the next big thing.

Brenner remembered years afterwards: 'In late 1962, Francis Crick and I . . . felt very strongly that most of the classical problems of molecular biology had been solved and that the future lay in tackling more complex biological problems. I remember that we decided against working on animal viruses, on the structure of ribosomes, on membranes, and other similar trivial problems in molecular biology. I had come to believe that most of molecular biology had become inevitable.'†

The choices facing him were like the difference between learning to read from a newspaper and just buying another paper. The cracking of the genetic code was a feat even

---

* E.O. Wilson, *Naturalist* (New York, 1994), p. 219
† Foreword to *C. elegans I*.

more astonishing than learning to read from an upside-down newspaper, and Brenner did not want to spend the rest of his life reading fresh editions of the same old news.

What remained to be discovered which would open up new worlds just as learning to read from a newspaper had done? It seemed that there were the two great problems of traditional biology, which should now be reduced to a molecular understanding. There was the problem of development, and the problem of behaviour.

Development is another of those things that appear unremarkable only because we so utterly fail to think about them. A single fertilised egg is round and apparently symmetrical in all directions. Yet when it divides the daughter cells form a distinct shape, with a front and a back, a top and a bottom, and the new cells at these edges become more and more specialised: they move and slide round each other; some die, and others elongate or swell, until finally you get an animal which is recognisable, complete, and made of very specialised cells all in their right places. If you take a cell from your heart, a cell from your liver and a cell from your fingernail, they will all have the same DNA, which contains recipes describing the proteins that make all three kinds of cell. But somehow each kind of cell knows, when it splits, that it must make a copy of itself, with only the proper bits of DNA expressed.* You don't get muscles on your fingernails, and you don't get bits of fingernail growing in your liver. This is good, but very odd, too: somehow the genome must contain a list not just of all the proteins needed to build a bit of liver, or a fingernail, but also of the proteins which manage to say, 'If you're in a liver, do only liver-cell things; if all your neighbours are fingernail cells,

---

* This knowledge, coded within cells, is what makes cloning mammals so difficult.

become one yourself.' What are these? How do they work? Do they in fact say nothing about where a cell finds itself, but only refer to its parents: 'If your mother was a fingernail cell, be one yourself'? Once you realise that these questions may have precise chemical answers which could be shown to grow out of the structure of the molecules involved, in the way that was true of haemoglobin, the universe seems to expand just as it did when you understood the sun was not just a bright splodge in the sheltering sky but a star, unimaginably distant and huge.

On top of development comes the problem of behaviour. Animals have nervous systems. They can move, and interact with the world, in ways which are clearly also read off their genes. This means that there must be genes for behaviour, just as there are genes for growth and development. Finding those and understanding how they worked would be a project worthy of a truly great scientist.

That is why Crick ended up as a neuroscientist. Another of the heroes of molecular biology, Seymour Benzer, helped by Brenner, had used a patient and extraordinarily complex programme of mutation to discover everything that could go wrong with a particular phage: first decomposing it into single genes, and finally breaking the genes he had discovered into single base pairs. Benzer now moved on to the fruit fly, an animal geneticists have used and possibly loved since 1910. His programme there was to discover the genes that govern its behaviour. Eventually his lab started to discover things like the single gene that seems to govern an animal's perception of time, so that flies in which it is mutated keep days longer or shorter than twenty-four hours.

Brenner started fooling around with bacteria, trying to analyse temperature-sensitive mutants which would function perfectly well at blood heat but which broke down in interesting ways when reared at 44°. It may seem vanishingly improbable that a mutant bacterium should appear in which everything works normally except that the cell wall isn't made properly at high temperatures, but the beauty of bacterial genetics is that the creatures multiply so fast that even a mutant which appears only once in ten billion attempts is bound to show up in a lab which is looking for it.* Brenner's programme with these bacteria was the same, in essence, as Benzer's successful attempt to find all the genes in a much simpler virus which ate bacteria had been. He tried to generate every possible mutation which disrupted the creature's growth until he had found – by breaking them individually – every single step needed to make it function right. One thinks of genetics as being about breeding animals to some constructive end; but, for this kind of biology, genetics is an essentially destructive exercise. It is an attempt to find every single way in which normal function can be broken in order to discover what are the bits it can be broken into, because these irreducible bits must represent the genes themselves. It's a little like playing the logic game Mastermind, where you try to guess a hidden pattern by analysing all the mistakes you have previously made.

This kind of Mastermind is how – for example – it was discovered (first in bacteria) that, while most genes code for proteins, some of these proteins are useful because they fit back onto the string of DNA, exposing or blocking bits of it, which lets them work as switches, turning other genes on and off. Just as in Mastermind, the discovery was worked out and proved logically necessary before the physical pattern was revealed. All these deductions about how genes

---

* Of course, this willingness to mutate is also why we catch fresh colds every winter.

worked were made decades before even the simplest gene or stretch of DNA had actually been analysed chemically and spelled out.

Brenner's next project was at least as bold. For what he proposed to do, in the summer of 1963, was to apply these analytical techniques to a real animal, something far more complicated than a bacterium. Having successfully reduced biology to the one-dimensional string of information contained in the sequence of bases along a strand of DNA, he wanted to return to three dimensions. As a young man studying anatomy in South Africa he had made a model of a bush-baby's brain by slicing it up very finely, copying each slice in wax, and then reassembling all these slices to make a three-dimensional model which could be opened up and examined at any point. Now he wanted to produce the same sort of model but at a much higher resolution; and the purpose this time was to understand how the animal built its three-dimensional self from the one-dimensional string of DNA.

Brenner was in a very strong position when he started the worm project. Any list his peers then drew up of molecular biologists who had already earned a Nobel prize would have had his name on it, and he could have had a job anywhere he wanted. But the LMB was anxious to keep him. By agreeing to his project, and to an extension building where it would later be housed, they were giving him plenty to do, and a degree of independence from Francis Crick, his collaborator, friend, and, unfortunately, superior in the hierarchy.*

* S. de Chadarevian, 1998. 'Of worms and programmes: "Caenorhabditis elegans" and the study of development', *Studies in Hist. and Phil. of Biol. and Biomed. Sci.* 29: 81–105 (also on web).

In summer 1963, he wrote to Max Perutz, who by then ran the LMB: 'I have long felt that the future of molecular biology lies in the extension of research to other fields of biology, notably development and the nervous system. This is not an original thought because, as you well know, many other molecular biologists are thinking in the same way. The great difficulty about these fields is that the nature of the problem has not yet been clearly defined, and hence the right experimental approach is not known.'

At first he looked at rotifers, tiny animals which live in water. But they didn't have sex often enough for the programme he had in mind: though they divide often and grow quickly, this is usually the result of parthenogenesis, with no genes exchanged in the process. 'This is an organism that treats the male sex with great disrespect.' *Naegleria*, an amoeba, was eventually dismissed as a member of the class of animals known by Brenner as neither vertebrate nor invertebrate, but 'pervertebrates'. Similar problems applied to *Tetrahymena*, a single-celled amoeba-like creature which looks like a minute green, hairy teardrop.* It could be grown in great numbers in water – Brenner's ideal organism would have been something fixed in place, at least at one end† – and it had been successfully mutated. But he did not like the genetics of *Tetrahymena*.

Flies had no problem with their sex lives, but they were too big and their lives too full of interest and excitement. Life in three dimensions would surely require a nervous system which was far too complicated to analyse from scratch. He needed something small which lived in a flat world, so he started to read up on nematode worms. A lot

* For a quick overview try http://www.lifesci.ucsb.edu/~genome/ Tetrahymena/Credits.htm.
† John Sulston, interview with the author.

of pioneering research had been done into the nervous system of a large parasitic nematode, *Ascaris suum*, in the nineteenth century. German biologists had studied the way cells divide as the animal grows, and showed that it appeared to be invariant: not just a simple proliferation of cells, but something that happened according to a definite pattern. Brenner could see from his vantage point in 1963 that this must be under genetic control and so might be analysed by the methods developed earlier with bacteria and their phages. *Ascaris* also had nerve cells big enough to see with an ordinary microscope and it appeared that these, too, were fixed in a pattern: each animal had exactly 298, which meant there must be genes to find that controlled the process. But *Ascaris* was about a hundred times too big for twentieth-century technology. He wanted something which could be sliced into pieces that would fit under an electron microscope; which bred quickly and sometimes at least sexually; and which thrived in laboratory conditions.

By late 1963 he had the right plan but the wrong worm. He wrote a grant proposal, on a single sheet of paper, to the Medical Research Council (MRC), the quango that was funding his research, on the day before they were due to discuss an extension doubling the size of the laboratory:

'We want a multicellular organism which has a short life cycle, can be easily cultivated, and is small enough to be handled in large numbers, like a micro-organism. It should have relatively few cells, so that exhaustive studies of lineage and patterns can be made, and should be amenable to genetic analysis.

'We think we have a good candidate in the form of a small nematode worm, *Caenorhabditis briggsae*, which has the following properties. It is a self-fertilising hermaphrodite, and sexual propagation is therefore independent of population size. Males are also found (0.1%), which can fertilise the hermaphrodites, allowing stocks to be constructed by genetic crosses. Each worm lays up to 200 eggs which hatch ... in twelve hours, producing larvae 80 microns in length. These larvae grow to a length of 1mm in three and a half days, and reach sexual maturity.

'To start with we propose to identify every cell in the worm and trace lineages. We shall also investigate the constancy of development and study its control by looking for mutants.'

There are several things about this proposal which seem astonishing with hindsight. The obvious oddity is that the strategy set out in the last two sentences, almost as a throwaway, took more than twenty years to carry out. But that is not what makes scientists giggle about it today. They are used to projects which grow unmanageably; and, besides, they know this one turned out all right in the end.

No, the one thing that everyone who remembers this paper pointed out to me was that it was a grant application on one piece of A4, asking for, and getting, a quite openended commitment to fund largely unspecified research. It is the sort of thing that simply could not happen today, but it is also an illustration of the style that made Cambridge such an effective centre for molecular biology in the Fifties and Sixties.

The world's centre of scientific gravity had not then shifted decisively to America. Molecular biology was done in small, cramped labs, and in large, spacious human brains but it didn't require very much high technology. Some experiments used radioactivity, and a lot depended on the development of very high-speed centrifuges, which allowed very small quantities of chemicals of interest to be isolated, but the most important computational technology involved was a blackboard and a first-rate brain. The centre of all this excellence was the Laboratory of Molecular Biology (LMB), distributed in odd corners of all sorts of buildings all over Cambridge, funded by the Medical Research Council and run by Max Perutz in a spirit of humble elitism. Ten Nobel prizes were won for work done there between 1945 and 2002.

The buildings of the LMB got grander over the years; and I suppose you could say that its final incarnation is the magnificent Genome Centre, named after Fred Sanger, which stands in parkland in the grounds of a small, pleasant eighteenth-century mansion south of Cambridge, which is now used as a conference centre. Sanger himself was once observed standing on the steps of the huge lab complex that bears his name, looking with disgust at the conference centre that is all that remains of the pre-industrial age there. 'Waste of space. They should put labs in there instead,' he said.*

The point about the LMB was that it did not believe in bureaucracy; it didn't even believe in desks. 'Desks encourage timewasting' was one motto, so people worked at cramped sections of shared benches. Equipment spilled out into the corridors. But the LMB did believe in constant conversation and hard work. Above all, it ran on the confidence that you would not find smarter or more hard-working people anywhere in the world, and the belief that if smart

---

* Eileen Southgate, interview with the author; compare Sydney Brenner's advice to the director of any research institution, from *Loose Ends* (London, 1997), a collection of his witty, but serious essays: 'When you get tired of arguing with your senior colleagues, you should have a talk with the person who looks after the refectory. You will find a refreshingly different view of the future, which is that you should get rid of all the scientists and their messy ways and turn the place into a first-class restaurant and conference centre.'

and hard-working people thought a problem interesting they should be given a little money and left to get on with it. Progress was measured by the opinions of their peers, not by anything easier to squeeze into a spreadsheet.

Modern molecular biologists can make vastly greater quantities of money than anyone at the LMB dreamed of, and they dispose of vastly greater budgets. But none, I think, can possibly have the freedom to follow their own judgment, and to work alongside colleagues doing exactly the same thing, that was granted to the original workers in the LMB. Brenner and his first helpers worked on the worm for five years without publishing a single paper, and for eight before producing something of which the world would have to take notice. That would be impossible in science today; yet without that freedom the project would never have happened, and quite probably the human genome would not yet have been sequenced either, since the people who did that had all learned their trade on the worm.

The informal and personal style of the LMB was typified by the process that brought Nichol Thomson to the lab. He was a Scot, a crofter's son from the tiny isle of Cumbraes in the Firth of Clyde, where the Scottish Marine Biological Association maintained a laboratory. Lord Rothschild, scientist and adviser to governments, used to come there twice a year to work on the fertilisation of sea urchins; and when Nichol Thomson married Ann Rafferty, who did secretarial work at the laboratory, Lord Rothschild brought them both down to Cambridge. Ann worked on his secretarial staff, eventually becoming his PA, and Nichol was trained as a lab technician.

Rothschild wanted to look at sea-urchin sperm under an electron microscope, then very new and rare; at first they had to travel to the physics unit at the Cavendish labs to use one. So Nichol Thomson became one of the first men in

England to master the crafts, perhaps the dark arts, of preparing specimens for the electron microscope, and soon one of the best at getting pictures out of it. One of the things I came to understand about the worm project was that when scientists say they couldn't have done the work without their technicians, they are entirely honest and often sincere. One thinks of scientists as being distinguished by their precision of thought, but this is useless unless it is matched by precision of experiment.

By the late Sixties, Thomson was so good at preparing worms for the microscope that it became feasible to want to model it by making 20,000 slices through this creature, which is only a millimetre long. The skill and dedication involved in such work was at least as great as that of anyone else in the project. Sydney thought Nichol Thomson 'a very talented lad' and was very keen to work with him; it's typical of the style of the MRC that when I asked Thomson whether anyone else in the world could have done his work, he replied, 'I don't know,' in a tone suggesting the question was really difficult to answer.

In 1964, Rothschild gave up active science for administration. He recommended Nichol Thomson to Sydney Brenner, and the great nematode hunt began in earnest. Brenner's first proposal had been for *C. briggsae*, a close relative of *elegans*, but very little was then known about small transparent worms, and everyone Brenner knew was set to collecting possible subjects for research. Even his children joined in, collecting wholly useless earthworms, but not discouraged from this. Nichol and Ann Thomson had a clearer idea of what was wanted. 'We'd get them from compost heaps. Even when we went to Scotland on holiday, we brought back – we took petri plates with us and brought back whatever we could find. Sydney claimed he always knew the ones from Scotland because they wore a kilt!'

The duty of nematode hunting was also laid on Muriel Wigby, who had been Brenner's technician since he arrived at the LMB. A quiet, shrewd woman, she had started work at eighteen: she came from a background where even clever girls did not go to university, and the work she did at the LMB demanded intelligence as well as care. She had a firm opinion of her own importance: she remembers a discussion in the lab when Francis Crick was speculating about how the statues on Easter Island had been brought to their final positions and raised in place. 'I said, "I know how they moved them. They stood around with their hands on their hips, talking, and thousands of technicians did the work!" I liked to undercut him like that. He was very pompous,' she explains. She had worked with Brenner on phage genetics, and then on bacteria before they moved to worms: eventually, after years of worm genetics, she retired, and bred dogs for pleasure.

On a sabbatical in Berkeley, Brenner met Ellsworth Dougherty, a medical doctor who also did research on nutrition, and who had a post in the university's zoological department. Dougherty had been promoting nematodes as research organisms at least since 1948, and he persuaded Brenner that *elegans*, rather than *briggsae*, might be the animal he was looking for. Back in Cambridge, Brenner dug some out of his own garden, and liked what he saw. The descendants of these worms were known as the N1 strain, but the worms on which the entire project was based, the N2 strain, came from Dougherty, who had learned to grow worms reliably on plates of bacteria and who had found in them the kind of mutation that Brenner knew he would be looking for: one which established the existence of a single gene controlling how well the worm would tolerate heat.

So by 1965 the project was established, even if only three people were involved. With the help of Nichol Thomson

and Muriel Wigby, Brenner worked at a small bench in the
LMB building just across from Addenbrooke's Hospital,
growing worms in slightly toxic environments to encourage
them to mutate, watching the results through the microscope, day after day, and trying to find ones which were
interestingly broken. Muriel Wigby found the first one,
'E1', which was later known as 'Dumpy'. Brenner found the
second, 'E2', which moved oddly. The work involved grinding monotony. A single hermaphrodite worm can produce
100,000 descendants in ten days, and it takes a rare passion
to look through 100,000 worms, trying to find the ones
that move differently from the others, to pick these mutants
out on the end of a wooden toothpick made sticky with
bacteria, and then to do so again and again with the
100,000 descendants of each of the original 100,000. Most
of his former colleagues found his change of direction mys-
tifying. Here was a man who had been swimming the
deepest and least-charted seas of thought in modern biology, setting off on an apparently endless trudge across a
desert of routine. Jim Watson thought he was twenty years
too ambitious. But Brenner was entirely happy. There was
nothing, he thought, that the worm could not accomplish in
the end: with the worm, you could discover that 'With a
few toothpicks, some petri dishes, and a microscope, you
can open the door to all of biology.'*

* Foreword to *C. elegans II* (Cold Spring Harbor Labs Press,
1997); also online at http://www.ncbi.nlm.nih.gov

2

The Worm

Caenorhabditis elegans is about the most unremarkable nematode known to man. There must be others, even less obvious, lurking undiscovered in the crannies of the world since nematodes are overwhelmingly the most numerous animals on earth. If a Martian biologist were to collect, at random, five million animals from the earth, sampling everything he could find, from apes and penguins through fish to the uncountable myriads of insects, almost all of them (four million) would be nematodes. The best estimates of the number of their species range between 100,000 and 10,000,000. The wild disparity of these estimates means that for every species we know about there may be a hundred that haven't been discovered yet, or there may be only ten. If Brenner had been interested in an economically valuable organism, he might have picked any of the numerous nematode parasites that cause humans suffering, either directly – about half the world's population is afflicted with parasitic nematodes, which can cause some extremely unpleasant diseases – or indirectly, because they parasitise almost everything humans eat: not just sheep and cattle, but plants ranging from coffee to carrots. Victorian biologists catalogued nematode parasites of lions, vultures, and seal's kidneys, these last growing up to 40 inches long. There is a gruesome saying among worm researchers that if everything on earth were to disappear except the nematodes, the outline of all plants and animals would be left, filled out by their nematode parasites. Just how close this comes to literal truth emerges from the fact that three different species of nematode can live in the rectum of the American cockroach, *Periplaneta americana*. There appears to be nowhere these animals will not try to make their homes.

Yet *elegans* lives in tranquil obscurity underground, parasitising nothing, eating only bacteria and slime moulds. In fact, hardly anything is known about its life in the wild. It appears to be very widespread: colonies have been found all over Europe, North America and Australia. There are now frozen collections, for reference purposes, of worms descended from ones captured in 1972 in a pineapple field in Hawaii, and in a flowerbed on the Caltech campus, among other places. But almost all the trillions of worms used in experiments since the 1970s are descended from a single colony found in a tray of mushroom compost in Bristol in 1954 by a man named Staniland, who passed them to the English biologist Warwick Nicholas. Nicholas, now living and still working in Australia, managed to get the worms growing in a purified (axenic) medium, and took a couple of sealed tubes with him to California, where he went to study with Ellsworth Dougherty. He did manage to established one mutant line of worms, but never got any further.

Dougherty was decades ahead of his time, and his fate was unusually horrible. Shortly after meeting Brenner, in 1964, he lost his post at Berkeley, largely as a result of political intrigues, according to Warwick Nicholas. The shock destroyed him. He sent Brenner, as promised, a culture of *elegans* in a couple of test tubes; but shortly thereafter, just before he was due to attend a meeting of the American Association for the Advancement of Science, he killed himself. He deserves to be remembered, not just because he was right beyond his wildest dreams in his ideas of what nematodes could do for human knowledge, but because the rest of this book is concerned with people who have been fulfilled and happy with their lives as scientists. Dougherty, perhaps consumed as much as any of those people by a passion for science, found it unrequited. There must have been others like him, and their ghosts writhe round the edges of this story, even if their failure was much less spectacular and they went on to live long, unscientific lives.

As a piece of social history, the worm project is a remarkable story of altruism, co-operation and general niceness. This may be in part because history is written and even remembered by the winners; but it is undeniable that there are lots of winners in this story and no conspicuous losers. Not everyone who worked for Brenner came to a happy ending, but it seems that most of them did. They gambled and they won. Dougherty's fate is a reminder of just how much a scientist risks who becomes unfashionable or unlucky in his gambles: yet no career was ever made without an element of risk.

C. elegans was preferred to the original choice, *C. briggsae*, because it was easier to breed and grow in the laboratory. Dougherty had tried for years to breed stable mutant strains which could not digest some nutrients, and so would be dependent on researchers feeding them particular amino acids. Such 'auxotrophic' mutants are very popular with researchers into bacteria because they are so easily controlled. You can use the food dependencies as labels for other, more subtle changes, so that an experiment can be set up in which all the bacteria with a particular mutation are also unable to synthesise a particular amino acid. If they are mixed with a population dependent on the supply of a different amino acid, you can at any moment sort the two by removing from their nourishing broth the amino acid on which one depends. The only ones remaining will be those which can synthesise the substance you have stopped supplying. It's a whole lot easier than counting them.

The worms in this story live on a diet of *Escherichia coli* bacteria modified to depend on the amino acid uracil: by limiting the amount of uracil available, the researchers can stop the bacteria growing so fast that the worms are buried in their own food, as they would like to be. But it proved very difficult to establish worms with nutritional deficiencies like that. And the advantages of *elegans* over *briggsae* turned out to be more subtle. The two worms appear to be close relatives. But close analysis of their DNA shows that they are in fact as distant from each other genetically as an elephant is from a mouse, but have converged on similar lifestyles and sizes, both being tiny transparent hermaphrodites, and even trained biologists were fooled by their similarity for decades.

It is a bit misleading to call either of them worms, because the word makes one think of earthworms, which are opaque and segmented, whereas nematodes are never segmented and often transparent. They are built almost like bicycle tyres, of two tubes, one inside the other. The outer tube is a tough skin, or cuticle. The inner tube contains the mouth, digestive tube, sex organs, and anus. They have no bones, no brains, no hearts, nor even any blood. Yet they have aims in life, and even behaviour that is sometimes disconcertingly human. If you feed a worm Prozac, it will wiggle less and eat more. It will also lay eggs, which a human under the same stimulus does not.

Elegans could more easily than *briggsae* be induced to grow on an agar jelly covered in nourishing bacteria – usually the same *E. coli* as had been used for the earlier experiments in phage genetics – and after a while Nichol Thomson discovered how they could be treated so they could be sliced thinly enough to fit under an electron microscope. That, says Brenner, was the clinching discovery. 'Nematodes were notorious in the literature for being impossible to fix and embed.' They discovered one reason for this impossibility by accident. One of the odder things about the worm is that, if it is starved while growing, it goes into *dauer*, a state in which it moves as little as possible and does not eat at all. Before Brenner and Thomson started their experiments, the worms that had been studied in the laboratories were trapped in the *dauer* phase, and when they are like that they do not absorb the fixative chemical necessary to prepare them to be cut into the shavings that an electron microscope will accept.

Brenner was working with so many organisms at the time that he left a plate of worms out until the animals had emerged from *dauer*; the results were fine. He had found his organism.

This preliminary work was a lot harder and more complex than it may seem. Brenner tried around sixty nematodes before he settled on *C. elegans*; and even after he had done so his success was hard to reproduce. François Jacob was another of the great pioneers of molecular biology, who with Jacques Monod had won a Nobel prize for showing that genes can switch each other on and off. He, too, tried to make a study of the worm in the late Sixties, but he could not get it to grow reliably in the laboratory. He turned to the mouse instead, but failed to get the funding for an Institut de la Souris.* It is not the smallest advantage of worms that they take up so little space and eat so cheaply. You need a great deal more money, space and time to run an institute for the study of mice. In fact, in summer 2002 the mouse storage faculties of the scientific world were almost exhausted:† in Houston alone, 40,000 newly built cages, each holding four or five mice, were full; and it was calculated that a really comprehensive investigation of the genome would require something like 60 million mice. Another advantage of the worm, though this could not have been foreseen in the Sixties, is that it has never fired the passions of animal-rights protesters. No one ever made a cuddly photograph of a nematode.

Worms grow through four larval stages between their emergence from the egg and adulthood.‡ At each moult they shuck their skins, rather as insects do. But the relationship between nematodes and any other family of animals is controversial. The nematodes have been around since before the cambrian explosion 650 million years ago. *C. elegans* itself has probably existed for 20 million years, but it's hard to tell with an animal which doesn't fossilise or leave much trace on history.

Little is known about the life of *elegans* in the wild. It has some parasites: there is a bacterium which lives inside its rectum and which sometimes finds its way into worm labs. Afflicted worms grow so constipated that they swell up, and they were at first mistaken for a new, exciting mutation. When the bacterial plates in laboratories are contaminated

* Rachel Ankeny, 'The Natural History of *C. elegans* Research', *Nature Reviews Genetics* 2: 474–8.
† 'Animal facilities are overflowing with mutant mice': *Nature*, vol. 417, p. 785.
‡ All of which look pretty much like worms.

with fungi, worms have been seen to climb up the stalks and shake gently, no one knows why. This is one of the few times when a lab worm can move in three dimensions. In the wild, they can move their heads in the air (the muscles in their bodies will only move them from side to side), but on plates of agar they are trapped by the surface tension for most of the time. Even so, they are sometimes observed to lift their heads and shake them in the air. Again, no one knows why.

When he was first watching the worms, Brenner coined the term 'dinosaur detector' to account for the fact that no one may ever find out, because parts of a worm's behaviour may be conserved reactions to threats which have long since disappeared from the environment.* Other parts may simply be adaptations to things which don't happen in the labs, where worms are only ever fed bacteria. They eat slime moulds in the wild when they can get at them, and it may be this requires reflexes which haven't been investigated yet.† Another possibility is that the N2 strain, which has become the standard for investigating the worms, itself carries a couple of mutations. There is some evidence for this: most worm strains collected from the wild burrow more often and more enthusiastically than N2 worms (which also makes them less suitable for examination under a microscope).‡ Many also have a slight sexual perversion unknown in N2 worms.**

---

* It's true that it's hard to think of anything *C. elegans* could do to protect itself from a dinosaur, but the principle is valid.
† Jonathan Hodgkin, interview with the author.
‡ Jonathan Hodgkin, *Worm Breeders' Gazette*, 8/3: 36.
** The males deposit a blob over the entrance of the vulva of a hermaphrodite after mating, though this serves no known purpose: it ought to stop other males from mating, but it doesn't.

Most of the life of a worm is spent eating, something at which it is fantastically efficient. Fifty per cent of the weight of bacteria it eats is converted to worm tissue, and its mechanisms for doing so are brutally efficient. The mouth cavity of *elegans*, seen under a scanning electron microscope, has around it six diamond-shaped sensory bumps which compress the opening into something shaped like a star of David. There is a photograph showing this head, full of vast and formless menace at 8,400× magnification, with a few hapless bacteria dangling from the lips. The bumps contain tiny channels which the nematode uses to smell the world around it and to detect the chemicals released by other worms.

Being an invertebrate, *elegans* has no jaws or teeth. Instead it has a pharynx, a short, muscular tube with a round bulb or crushing chamber at the far end. Each section of the pharynx lies between three muscles running along its length. At rest, these muscles relax, constricting the tube between them into three folds like a Mercedes emblem in cross-section. When they contract, the sides of the pharynx spring apart to make a triangular tube, into which any liquid in front of the worm is sucked. When the muscles relax again, the liquid is expelled but any food particles it contains are trapped within the pharynx. It's odd to reflect that this is essentially the same method of feeding as is used by the largest animals on the planet, blue whales, which also suck their food in as liquid and then strain out the good bits. Whales, however, have visible strainers for their food. The worm has no strainers visible, even under an electron microscope. The separation may be accomplished by some exquisitely tuned turbulence in the liquid as it is pumped through the channel, which is much thinner than a human hair.

From the worm's pharynx, the filtered-out bacteria move backwards to the grinding bulb at the end. It is lined with knobbly projections which crush open the bacteria, so their nutrients pass backwards into the gut, propelled at high pressure by the pumping muscles of the pharynx. The interior of the worm is pressurised: if you prick one it does not bleed but it will burst. This means that it needs no muscles in the intestine. All it needs to do is relax the muscle closing off the other end and everything left from digestion is expelled at the anus: a healthy worm defecates about every forty-five seconds all its life.

This life, devoted to eating and breeding as much and as often as possible, is a reaction to the uncertainties of life in the soil. It is a classic example of what is known to biologists as *r*-selection: you have as many offspring as you can and care for them as little as possible in the hope that some will survive. The alternative strategy, *k*-selection, pursued by most mammals including us, is to have relatively few offspring, but to invest a great deal into each of them, thus increasing their individual chances of survival. The soil in which *C. elegans* lives is not a homogeneous environment. It has pockets of juicy bacteria, but long intervening stretches without much food. It is variously dry and damp: studies in the wild have shown that *C. elegans* moves within the films of moisture separating grains of earth. If these dry out, it is trapped by the surface tension and will die. *C. elegans* also has its predators. Earthworms eat them, and there are species of fungi which specialise in trapping nematodes and digesting them.

In fact, the greed of *C. elegans*, though effective in the short term, causes it long-term problems, because the animals devour all the available food as quickly as possible. Nor are they well adapted to long foraging voyages. The answer to these problems is supplied by the *dauer* state, which corresponds to the strange encystments practised by many parasitic nematodes, which can spend much of their lives quiescent in odd parts of an unwitting host, such as the lungs of a sheep or the eyeballs of a frog. As soon as a worm is hatched, the growing, feeding first-stage larva is sensitive to two chemicals in particular, which it samples or smells through the bumps around its mouth. One is a pheromone produced by all worms, with extremely long-lasting effects. This means that a growing worm can smell pretty accurately how many worms are sharing its patch of bacteria. The second chemical is the smell of food. If the smell of worm is stronger than the smell of food, this sends a signal that there won't be much to keep an adult alive. The growing larva can distinguish between different strains of bacteria, and modifies its choice according to the nutritional value of the available food. Just what constitutes 'stronger' is affected by temperature, too. The higher the temperature around a worm – they thrive best at around 20°C – the more likely it is to turn towards *dauer*.

Normally-developing larvae spend only seven hours in the second stage of their lives, and most of that time is spent growing their gonads and eggs. But a worm which senses too many other worms around it, and too little food, spends nearly twice as long as a second-stage larva, and in this time it lays up fat in the cells of its skin and intestine, rather than developing eggs. When next it moults, to become a level-three larva, the *dauer* is immediately recognisable because it is thinner and darker than a normal worm. Its stomach has closed up: in fact, the mouth is completely closed by a block of the cuticle that normally forms the worm's outer skin, and the pharynx is shrunken. Its skin acquires a protective, water-repellent coating. In this stage, the larvae can survive for up to three months, waiting for food to reappear. They are passive, but not inactive. Think of them as sulky adolescents, conserving their energy for some grand adventure which may never come. They move quickly if touched, and they seek out water: when left in petri dishes in the lab, they crawl at night into the droplets of condensation that form inside the lids and there form huddles. But mostly they just lie around the agar, waiting for fate, or food.

Perhaps the oddest *dauer* behaviour is nictation, when the larva climbs any projection it can find and balances on its tail, waving its head in the air. The only explanation proposed for this is that it is trying to hitch a ride on a passing insect or animal – imagine this happening in a tunnel in the soil – in the hope that it will be carried to fresh food supplies. If that sounds odd, it is not unusual by the extraordinary standards of some of *elegans*'s parasitic relatives. *Bursaphelenchus xylophilus*, a minute nematode which infests pine trees, normally lives in the resin channels of its host. But it sometimes forms a *dauer* stage in the pupal case of bark beetles, which also infest the trees, and when the insect hatches from its pupal stage the *dauer* larvae crawl into its breathing apparatus, where they are transported to fresh pine trees when the beetle flies off.

Heterorhabditis bacteriophora has an even more roundabout way to make its living by a mixture of symbiosis and parasitism. *H. bacteriophora*, like *C. elegans*, lives on bacteria, as its name suggests, but the bacteria that *H. bacteriophora* prefers to live, if they can, on insect flesh. So the *dauer* stage of *H. bacteriophora* has a special pocket in its gut, in which it carries spores of a peculiar bacterium, *Xenorhabdus*, which not only eats insect tissue but secretes an antibiotic which kills rival bacteria. Both these creatures are minute compared to the caterpillars they prey on: the *dauer* larvae of *H. Bacteriophora* crawl into their prey and then migrate to its body cavities, where they sick up or defecate the spores of their bacterium. These come to life and multiply enormously until they have consumed the entire insect from within; the nematode, meanwhile, emerges from *dauer* and starts to harvest the bacteria it has planted, and to breed. Once the whole animal has been consumed, the new-born worms form *dauer* larvae and move away to find new caterpillars. Such natural means of pest control are preferred to chemicals by many gardeners who pride themselves on their sensitivity.

None of the animals involved in these extraordinary patterns of behaviour has a brain, of course.* What they do is just a consequence of the electrical and chemical traffic in their nervous systems, and if all that traffic could be traced in its molecular detail we would have a complete explanation of the worm's behaviour. That was the grand design, at any rate.

Dauer larvae survive high temperatures much better than adults. At 37°C (human blood heat), they live three times longer than adults, and, whereas the adults crawl around to try and stay in the temperature at which they were hatched, the *dauer* actively seek out new temperatures. This makes sense as foraging behaviour, trying to explore where no worm has gone before.

The worm recovers from *dauer* under the opposite stimuli to those that sent it away. Given food, low temperature, and a weak smell of other worms around it, its throat and skin return to normal and within three hours the pharynx is pumping away as the worm feeds normally. Then it moults into its final larval stage, and after one more moult is ready to breed.

The most important thing about C. *elegans* at the beginning, apart from the fact that it could be displayed under an electron microscope in illuminating ways, was its sex life. It has sex early and often, usually with itself. These two facts

* Except perhaps the gardeners.

make it fascinating for geneticists. *C. elegans* takes less than four days to grow from an egg to an egg-laying animal, so the results of an experiment are quick to appear. The sex organs of a hermaphrodite *C. elegans* take up most of the middle of its body. They resemble an art nouveau 'y' which has been squashed flat, so that the very short stalk is the vulva, and the two long arms, each folded back once on itself in a hairpin bend, are the gonads, where eggs and sperm develop. The eggs start to develop at the far point of the hairpin, and gradually grow as they move round the bend in the tube towards the vulva. The sperm, which are repulsive to look at,* lurk at the exit of this tube, and fertilise the eggs before they move into the vulva. Unlike most animal sperm, those of nematodes have no tail and don't wiggle or swim; instead, they are amoeboid blobs, which drag themselves along a surface by expanding and contracting the cell walls like tiny caterpillar tracks.

Once fertilised, the embryo develops inside the egg until it has grown to a tiny larva, which hatches once the eggs have been expelled from the vulva. Since the worm and its eggs are transparent, this process can be watched through a microscope as it happens.

The mathematics of worm sex are mind-boggling. I once tried to work out how many worms had laid down their lives for science in the last thirty years, and decided very rapidly that the number was incalculable. I mean that quite literally, and not just because my spreadsheet refused to contemplate them. The beginning of the calculation is easy: a worm grows to maturity in about three and a half days. After that it starts laying eggs. Most lay about 300 eggs

---

* Words fail me here. But there are Quicktime movies of the moving sperm available on the web, at http://www.mcb.arizona.edu/wardlab/gallery.html.

over the next four days; each egg will hatch in four days' time into a hermaphrodite which will lay another 300 eggs. So one worm has 300 children and 90,000 grandchildren. The 90,000 would, if food were unlimited, produce 27 million children of their own. By the end of a month, assuming unlimited food and room, one worm could have 8,000 million living progeny, or, as an American would say, 8 billion. There aren't that many people alive on earth. After the second month, each of those 8 billion worms could have produced another 8 billion descendants, giving as the total number of possible descendants of one hermaphrodite in 69 days a figure which has 27 zeros after it.* No wonder the spreadsheet boggled. When God promised Abraham descendants who would outnumber the stars in the sky or the grains of sand on a beach, He didn't mention that He had made the same promise to a worm first. Almost all the worms in all the laboratories around the world today descend from a single hermaphrodite clone chosen by Brenner in 1965; its true-breeding descendants are known as the N2 strain. In this perspective it is absolutely astonishing how very few mutants have been found and how effective the machinery of DNA repair and copying must be

---

* The numbers grow absurd just as quickly if you look inside a worm. A single *C. elegans* is just about visible with the naked eye. It contains about 2,000 cells (including eggs and sperm), each of which is visible with a light microscope. Among the smallest and hardest to see of these cells are the developing eggs. If you study an egg through an electron microscope, magnifying it another thousand times, you can make out the chromosomes. Those chromosomes conceal 100 million base pairs of DNA, and those are the individual molecules for which the early worm project ended up searching. The human being, by the way, has thirty times as much DNA in each cell, though the cells are notably bigger.

to have produced so many worms which have nothing interesting about them at all.

Hermaphroditic sex means that each worm has two parents, and two sets of genes, even though both sets come from the same animal. One nice effect of this from the scientist's point of view is that there can be no hidden recessive genes in *C. elegans*. In any hermaphrodite carrying a particular mutation, half the eggs and half the sperm will also carry a copy of it. This means that about a quarter of the progeny will have the mutation from both sides of the parent, and so will express it for certain. Since there will be at least a couple of hundred eggs, you may be certain of finding every recessive mutation quickly, since fifty or more worms in every generation will show it, unless its appearance is masked by some other unrelated gene. It may take a few generations for such mutations to become visible: many of the methods used by Brenner in the early days relied on disrupting the development of the egg or the sperm cells and so the mutations did not appear for two or three generations after the poison was applied.

The advantages of a hermaphroditism were only half of what made the worm attractive to the geneticist in Brenner. Hermaphroditic sex makes it easy to isolate and identify mutants, but a complete understanding would make clear the relationships between the various mutations. To do this, you need males. The male *C. elegans* is produced by a chromosomal accident. The hermaphrodite has five pairs of A chromosomes, and two X ones. About one time in a thousand, when the chromosomes in the parent are being sorted into eggs and sperm, by first doubling and then splitting, one of the doubled Xs fails to separate, leaving an animal which has only one X chromosome, a male. In male worms the gonad develops as a single tube running down towards an exit on the tail. On the end of the tail, round this exit, there are nine tiny palps, like splayed-out toes, and a pro-jection like a tiny thorn, called a spicule. These are used to find the vulva when the male finds a female and then to anchor there.

When a male approaches a female it touches her nor-mally on the back, and then performs a delicious writhing slither all round the tail until it can insert the spicule into the vulva; one of the oddest sights I came across while researching this book was a roomful of worm biologists transfixed by a (literally) blue movie showing the sexual misadventures of two three-foot-long worms, in which the male had had a crucial neuron burned away with a laser beam, and so had ludicrous trouble finding the right hole.

But in the normal course of events the males mate very efficiently. Though we can't tell the difference between male sperm and hermaphrodite sperm, worms can, and a hermaphrodite will always fertilise its eggs with male sperm if any is available. Even if a hermaphrodite has used up all its own sperm, it may still have several hundred unfertilised eggs to lay, and a male can fertilise them. Given sufficient access to hermaphrodites, a single male can have nearly 3,000 progeny.

Why this matters to geneticists is indirect: they have no particular interest in healthy males, but if a hermaphrodite with an interesting mutation can be induced to have male children, these will also carry the broken gene and can be used to transfer the mutation to another hermaphrodite. Though males are very rare in wild-type worm populations, a number of mutant strains have been established which carry a gene which ensures that males become 300 times more common than they should be, so the method is reli-able. Short of artificial insemination, which was for many years not practicable in worms, this was the only way to get genes from one hermaphrodite into another. But, since it works well, Brenner had the tools he needed to turn the worm into a giant genetic jigsaw puzzle. He could map both the physical relations of genes to each other along a chromosome, and their logical relationships – which ones interacted with which others – and he could do all this by using males to put two genes of interest together into the same hermaphrodite.

The trick to the first kind of mapping is very simple. You put the two mutations together in a single worm, and then study how many of its descendants have both of them together. Genes can get separated from each other when the chromosomes on which they sit are split up and recombined in the formation of eggs and sperm. Genes on separate chromosomes altogether will obviously be split more often than genes on the same chromosome. Genes which sit close to each other on a chromosome will be split less often than those which are at opposite ends. Using these simple principles, it is possible to measure, over hundreds of generations, just how close to each other any particular pair of genes is. The work is tedious and exceptionally time-consuming: but when Brenner eventually published his findings on the genetics of the worm, eight years after settling on it as a research organism, he had established that the 550 genes he had found could almost all be sorted into six groups which travelled down the generations together, corresponding to the six chromosomes on which they could be found.

The second sort of gene mapping, logical mapping, is more subtle and more impressive. It is a way to find genes whose products interact with each other, even if they travel on different chromosomes; and it depends on an understanding of the chemistry of proteins.

Once a protein has folded up into its final shape, there will be patches where its knotted macaroni chains of amino acids match exactly other patches on the outside of other proteins. These 'binding sites', as they are known, are very specific. When you change the shape of a protein by changing its constituent chain of amino acids, you often change the binding site so that it no longer fits the other proteins as it used to. In fact, this kind of change is one of the commonest ways in which mutations in the DNA have their effect on the bodies around them.

Perhaps the most brilliant trick of the phage geneticists was to see that, once you had knocked out a protein in this way, another mutation in a different gene might change the shape of a protein which had fitted the first one before its distortion so that the two once more fitted together. Once that happened, you could still map the two genes physically, in the old way, by seeing how often and how soon they diverged in the descendants of animals which carried both mutations. But you would also have the beginnings of a logical map showing which genes produced proteins that interacted with each other in the economy of the cell.

The cleverness of this idea was that it required no detailed knowledge of the underlying DNA. When Brenner started the project, there was no reason to suppose he would ever have access to that sort of information. No one had yet cloned a gene, by making it grow inside an unrelated bacterium, and the idea that a chromosome could be directly read out, as happens today in the human genome project, was one for wild dreamers. But an enormous amount could be learned by these methods of indirection and careful breeding. Once the logical linkage between two genes had been established, it was possible to work up and down the chain, finding other genes affected at each end, until at last you had a complete picture of one of the little subroutines that cells are full of, in which nine or ten proteins make or break down each other in a long sequence of reactions until the final protein appears, which, among its other effects, fits into some of the first proteins involved, which shuts down the whole reaction in an elegant process of feedback. Actually discovering which proteins are involved, and how they fit together, would be the work of another lifetime, but, using these methods of genetic analysis, you will at least discover where to look.

The simplicity of the worm made all this possible. No animal with a brain could possibly be so gratifying to a geneticist, since brains register learning by physically changing the pattern of nerve cells and their connections. That means that you cannot easily and simply analyse the way genes specify a nervous system in more complex animals, because the way the nervous system ends up depends on the animal's experiences as it is growing. But with *C. elegans* it is always exactly the same. There is a rigid pattern of development, so it made sense to try to disrupt it in as many different ways as possible.

When Brenner finally published his catalogue of worm genetics, in 1974, he had identified over a hundred mutations which made the animal wriggle badly – he called them *uncs*, for uncoordinated. Some were caused by the nervous system growing wrongly; some by the nervous system failing to function once it had grown. The most productive class were caused by muscle defects, which gave him a way of seeing exactly which proteins worm muscles were made from. With his first collaborators he was able to show that his lab had learned to hack open the worm, not with scalpel blades but with genetics.

3

The Programme

Hacking open the worm in a new way rapidly became Brenner's passion. After two or three years making mutants, he discovered computer programming. As a young man in Johannesburg, he had shared an office for a while with Seymour Papert, who later went off to MIT and invented the Logo programming language for children. So he was aware that computers existed, but it was not until the late Sixties that they captured him. Nowadays he has his e-mail answered by a secretary, preferably one in another continent, and he mutters that young scientists get no work done because they spend too much time on their e-mail; but this abstention has in it something of an alcoholic's ostentatiously swigging ginger ale. His first feats in computing were extraordinary. To the people around him he seemed possessed by the machine, and he himself said once that for about eighteen months he more or less disappeared into the computer.

Brenner's first idea with the computer was that it would be able to store and display all the maps of the worm he was reconstructing from the slices that Nichol Thomson was photographing. When he had modelled a bush-baby's brain as a young student of anatomy, the resulting model was a physical lump he could hold in his hands. But the maps he needed of the worm would be so finely detailed that any physical model of them would be unmanageably huge.* So Brenner planned to virtualise the worm. Every slice would be photographed, and the photographs would be reassembled inside a computer to make a model which could be explored on screen from every angle.

Nowadays such a plan seems perfectly sensible. Indeed, twenty years after Sydney Brenner had the idea, you could have made yourself a name as a visionary in the computer world by calling it 'virtual reality'. But this was the late Sixties. Not only were the most powerful computers a lot less powerful than a modern mobile phone, but they were probably less powerful than even the stupidest washing machine you can buy today. They had few screens, no keyboards, no proper printers and at first no disk drives. At first, all they had to communicate with the outside world were long strings of paper tape with holes punched in it, like telex machines used.

The other problem that Brenner ran up against was metaphysical, and persists to this day: any computer you can buy is already obsolete and unable to do all you need, whereas anything powerful enough to do what you need is due in six months but doesn't exist today. Eventually he found an English device called the Modular 1, which was poised on the cusp of obsolescence and non-existence, and had it installed in the new MRC building, then being built at the end of the old one, partly to provide Brenner with a playground big enough to keep him in Cambridge.

In the Sixties computers were immense. The Modular 1

---

* They did try one small model; see p. 80.

took up most of the ground floor of the shell of the new building. 'It was the most wonderful building I've ever worked in. There were no services except electrical services, which we led in on open wires to the floor. So essentially we had this sort of vast hall with the computer sitting in the middle and acres of space to enjoy ourselves with. We needed acres of space because in those days the computer had no means of doing anything except an assembler program. There was no high-level language programmed for it and all input and output was done on paper tape. I can remember lying on the floor with tons of paper tape unwound, editing individual punched holes.'* The object of this veneration had 64k of memory,† but came with a display capable of quite advanced graphics if they first wrote and debugged all the display software on paper tape.

So it came about that the second man hired on the worm project was not a biologist at all, but a computer graphics expert named John White, whose face and manner have something of the gaunt knowingness of Leonard Rossiter.

To be a computer person in the late Sixties was to be a misfit almost by definition, but White was not even a conventional computer scientist. He had grown up as one of those boys who are always building gadgets: 'bombs and rockets and radios, that sort of thing,' he says, as if the logical connection were obvious. But when he left school he just dropped out for a while, until he found himself working as a technician at the MRC laboratories in Mill Hill, north London. He ended up in an electronics lab, making gadgets for the rest of the lab, and this awoke an interest in biology.

* Brenner and Wolpert, *A Life in Science*, p. 62
† Four thousand times as much memory as that, mounted on a single chip which I can hang off a keyring, cost me £35.00 in summer 2002.

He was obviously bright, so the MRC encouraged him to finish his education with a 'sandwich' degree, whereby he worked part-time and took courses in six-month bursts.

Once he had secured his degree, he reckoned he had better things to do in life than make gadgets for other people. He was working on the ways in which nerve cells make connections and store memories, and at the same time beginning to dabble in computer graphics, but there were no research jobs available at the London MRC, so he applied for a job putting computers into telephone exchanges. Just as he was about to accept it, his boss told him that a man in Cambridge was interested in computer graphics, so he read up on DNA in *Scientific American*, put on his only suit, and went up for the interview, where he found Sydney Brenner chain-smoking in jeans and a Breton fisherman's shirt. 'He didn't stop talking, really. He talked to me all about these things he was going to do: cell biology, developmental biology, and how the nervous system works – all this sort of thing. It was really quite extraordinary. We shuffled around the labs and I was absolutely bombarded. As I left, he said, "Well, you know, there'd be no trouble about you transferring, because this is just another MRC appointment."'

Brenner's vision of the worm as biological Meccano caught both sides of White's imagination, the mechanical, tinkering one, and his growing interest in biology; genetics united them in the single idea of computation. You can look at a nervous system in two ways at least. John Sulston – the next man to join the project – was looking at the nervous system as a complicated chemical network like a giant factory. But you can also see it as a kind of computer. It is constantly taking inputs and turning them into behaviours according to rules which might be written out as logic. So discovering exactly how this biological computer is programmed was a perfect sort of science. The computation requires no brain, since the worm has none. Yet somehow it moves in ways that are mathematically predictable; and taking computation in the larger sense, the worm manages to shuffle its 20,000 genes and deal them out in just the ways that are needed to specify the growth of a whole animal. There must be programs there, of a sort about which human computer programming knows very little.

Three days after the interview, Brenner rang White and asked when he was starting. The whole thing was done so quickly that for some months the MRC central bureaucracy did not realise White had moved to Cambridge and made difficulties about paying him there. That was not the only surprise: he found himself in a world like nothing he had known or imagined before. 'It's difficult to portray the amount of energy and charisma that Sydney exuded. He would just talk incessantly about things, and his excitement was completely infectious; but he also worked most incredibly hard. Everything was done in cramped labs, with equipment spilling out into the corridors, and the place was full of post-docs who never seemed to sleep.'

Brenner's fantastic drive and energy were partly an expression of his temperament: David Hirsh, who worked at the MRC in the late Sixties, remembered that Brenner routinely worked until three in the morning, and started early, too. But they were also a recognition of the risks of the project, and of the fact that it would work only if everyone involved believed in it as much as he did, and worked as hard at it as he did. The project of completely understanding an animal even as simple as the worm was so ambitious it was almost impossible for anyone to grasp, unless they were armoured like so many of Brenner's early recruits by ignorance of what they were taking on; there was also the problem that, if you were going to try to get to a complete understanding of an animal, most biologists would not have started from *C. elegans*.

Brenner had the gift, necessary in great scientists – and, for that matter, great artists – of looking at a problem and reassembling it from the ground up, so that it made a new and altogether simpler puzzle which no one had seen before. But these reconstructions are most convincing to people who have never grown used to the old problem, and so can't imagine how difficult it ought to be to solve.

Brenner and White would lie on the floor beneath their computer like the sons of Laocöon, wrestling with snakes of paper tape, debugging programs by looking at the pattern of holes in the paper and when necessary punching new ones. The machine was 'the exact opposite of user-friendly. It found all users most distasteful and seemed to go out of its way to avoid any constructive interaction with us. Most of the time it responded by going absolutely silent; at other times it issued cryptic remarks before collapsing completely.

'Our paper-tape reader sometimes had outbursts of hysterical rage and the tape would emerge neatly shredded into several strands.'*

The first useful thing Brenner wrote was a program to help analyse fragments of tRNA (tRNA helps to assemble proteins within the cell); but this would have been a tremendous labour in assembly language. He found in a journal a description of another programming language to write the sequencing program with, so he wrote, in assembler, an interpreter for this second language, TRAC. TRAC, the string-processing language turned out to be useful in writing the text editor they needed if they were to write the operating system extension to read the disks they were going to

---

* Brenner, *Loose Ends*, p. 42

need to store all the data when that started coming in . . . In the end the whole project grew with almost as much disorganised vigour as if it were alive. Finally, he was able to use it to write the sequence-analysing programs he had first thought of.

The work with TRAC makes Brenner the first man to identify and solve a problem which twenty years later was to be crucial to the sequencing of the human genome. The billions of bases in a genome sequence are almost impossible to sort through. To find the particular chemical of interest is not like looking for a needle in a haystack: it's like looking for a haystraw in a haystack. Without computers it would be completely impossible. In the library of the Genome Centre at Hinxton, there is a magazine called *The Journal of* in Silico *Biology*, and that, too, descends from Sydney Brenner and John White lying on the floor beside their computer, wrestling with paper tape.

'He used to drive people mad because he would get so excited about his programs. He'd rush in and show people some new algorithm that he'd got working and they would look on with glazed eyes,' says White.

The programming was done early in the morning or late at night, after Brenner had spent long days working with his mutants. They were the base on which the whole project rested; ideally, he would have discovered every single gene in the worm by breeding them until he had found all the mutations that could possibly knock out a gene. Of course this isn't possible, even in theory, because there are plenty of changes in genes which leave no trace on the animal's growth or behaviour, or at least no trace that is ever observed in laboratory conditions. Such mutations are said in the jargon to have no phenotype. But, by steadily and systematically poisoning his worms just enough to encourage them to have malfunctioning offspring, Brenner was able after a while to produce thirty or forty mutants in every batch of three hundred eggs. Each mutant had to be watched while growing so that they could be identified.

Muriel Wigby did an enormous amount of the grunt work here, for Brenner was still working on phage as well as computers and the worm. What she remembers now is the tedium. 'The sheer hard work of it is actually what genetics is all about.' Many of the mutagenised worms died. Others were sterile. Others looked interestingly deformed, but had descendants which reverted to the wild type. Yet the researchers were able to find a sufficient number which not only grew up crooked but passed on their defects to their descendants, showing that their DNA had in fact been changed in interesting ways.

But that was only the beginning of their labours, because the same behaviour or appearance may be produced by numerous different mutations. Anything that we can see on the outside of an animal is the results of processes involving the products of many genes; when one gene goes wrong, it may have effects all over the place; and a single visible defect might be produced by the malfunction of any one of tens or hundreds of genes acting together.

Phil Anderson, a much later worm geneticist, put the whole programme succinctly. Imagine, he said, that you are trying to do car mechanics by the genetic method. 'If you take one operating automobile and now derive from that an automobile that doesn't go down the road, there'd be a lot of ways that you could impair function of that automobile. On the one hand, if you could take out the water pump, it wouldn't go very far. If you took out the fuel pump, it wouldn't go at all. If you took out the battery, you could never get it started. By investigating the behaviour of that automobile in the absence of single components, you're going to understand what those components do. Take out the battery, the car moves the same but you can't start it, but if you can ever get it started it runs just fine. The genetic method is really that. It works by subtracting single components, meaning genes and their encoded proteins, and asking what goes wrong.'

The resulting catalogue of worm defects has a strange poetry of its own. Flicking through the list of *unc* phenotypes, I found:

44 paralysed coiler, dumpy, tends to curl.

46 shrinker, contracts both dorsally and ventrally when prodded; slow, good forward movement, poor backing; adult male moderately constipated.

52 adults limp, paralysed except for head regions, thin; larvae move well with progressive dystrophy.

54 limp paralysed phenotype at all stages; larvae can move slightly more than adults. Muscle ultrastructure very disorganised.

42 slow, lazy, slightly rippling movement, poor backing, thin, almost paralysed, tends to shrink and relax when prodded.

35 loopy irregular forward movement, poor backing, active, slightly thin.

23 'benthead' phenotype: progressive dystrophy of head musculature so that the adult head is bent dorsally or ventrally.

26 severe kinker, small, scrawny, flaccid, little movement, some have an expulsion defect in defaecation.

4 large, healthy, active, moves forward well but cannot move back. Ventral cord VA neurons have normal anatomy, but most have synaptic inputs appropriate to VB motor neurons.

67 sluggish: can move well both forward and backwards but frequently pauses or jams up.

Each of these complicated patterns of defect is caused by the malfunctioning of a single gene. The real labour of Brenner's scheme lay in the patient isolation of each mutant by endless crosses to establish that different clones of worms exhibited the same quirks and failures as a result of different genetic errors. Distinguishing *unc*-46 ('shrinker, contracts both dorsally and ventrally when prodded; slow, good forward movement, poor backing; adult male moderately constipated') from *unc*-42 ('slow, lazy, slightly rippling movement, poor backing, thin, almost paralysed, tends to shrink and relax when prodded') cannot have been easy, and it is sobering to contemplate the amount of patient scrutiny that must have gone into noticing and then identifying *unc*-67, which looks normal and can move well both forwards and backwards 'but frequently pauses or jams up', on a plate full of worms all moving, and often pausing, as they crawl over their bacterial lawn.

Then each newly identified mutant had to be further tested, for touch sensitivity (you tapped them with an eyelash hair), tolerance of different temperatures, and drug resistance. 'These were still the Sixties, really,' said someone who was there: when testing for drug-resistant mutants, they used every drug they could get their hands on, legal or not. But the worms never did anything very interesting on cannabis or LSD, and one morning, so the story goes, the lab's stock of acid was found to have disappeared from the refrigerator. 'The worms did not consume it,' I was gravely told.

When dealing with bacteria, or their viruses, this kind of project is easier. In organisms that simple there really is one gene for every capability you can observe. The classic experiments of molecular biology involved genes which did very simple things: make an enzyme which lets a bacterium digest sugar; or make another protein which attaches somewhere to the DNA and switches off the first gene.

This kind of molecular switching looks an awful lot like the inner logic of computers, with their IFS and THENS and GOTOS. You can imagine the DNA as a long paper tape with the program for a bacterium on it, so that a fragment, written out in pseudocode might look like this: IF there is sugar, THEN make sugar digestion gene, ELSE make sugar-off gene. The ultimate dream of the early days with the Modular 1 was to find all those instructions within the worm, and write them out into one long program which would tell you exactly what a worm's DNA meant, and so what a worm must be.

In a way, this dream preceded the computer. Bob Edgar was a molecular biologist of the older generation, who had started his studies even before the structure of DNA was unravelled. 'When I went to school and did genetics, we didn't know what a gene was! It might be protein, it might be DNA, it might be RNA, it might be a complex of both, it might be nothing. And those early people that I worked with, like [Max] Delbrück and so on, I mean, they were mystical about all this. Theirs was a mystical quest.'

For him, the whole point of the gene was that it was an almost scripture, in which all the secrets of life were written. 'I was a first-year graduate student when the [Crick and Watson] paper came out in 1953 – I and my fellow graduate students, we were absolutely flabbergasted and incredibly excited about this; I remember immediately thinking, "The secret messages that are written in the DNA!" [We thought] the secret of life is somehow embedded in the genes and I was just absolutely giddy with excitement at the thought of being part of this process.'

He was not in the least bit interested in how they were written. It was the existence of the code that had captured him: the idea of a store of pure information at the heart of life. 'If you could read this genetic manual, you could just read it out, bing, bing, bing. It was not the fact that it was making proteins – none of that interested me. It was this idea of having the whole picture before you, a whole organism.'

Brenner was always more interested in the squishy details of biology. He tells with approval in his autobiography the story of a physicist who complained, when he switched to biology, that he could no longer have three-hour baths in which he did his thinking, because, as a biologist, he kept having to leap from the bath to look up a fact which might derail his lovely new theories. Yet he shared the dream that they might find within the worm's DNA its Rosetta Stone and that this could be replicated in a computer program.

On the way to this end – still unattained in 2002 – the two men ended up writing a complete operating system for their computer, entirely in assembly language. This is the kind of feat which today, on its own, makes the reputation of someone like Linus Torvalds, the author of Linux. White's casual reference to building radios and bombs as a child pales by comparison. But it was only the beginning of their labours. In fact, by the end they had written several operating systems for the one machine.* Once they had that done, and there was some chance of the computer actually becoming useful, their work got even closer to the machinery. Brenner spent some months writing what would now be called a printer driver, except that there weren't then any real printers to drive. All that existed were teletype machines, golfball typewriters which had a connection to a computer instead of a keyboard. One of them had to be modified so that it would print the results and comparisons

---

* Interview with the author.

of his nucleic acid sequencer over six pages of paper, which could then be stuck back together to give the big picture. White, meanwhile, worked on a mechanical device which would photograph slices of worms automatically; and the two men together planned a program which would scan the photographs and assemble from them a 3D map of the worms.

Of course, the computer was fun for everyone on the project. Jonathan Hodgkin, then a graduate student, remembers that it was a marvellous toy for 1971, when computers were hardly known. 'Having access to quite a nice computer, people who could tell me about how to do things and who'd say, "Obviously, you'd better teach yourself programming and here's this toy you can play with" – that was just wonderful.'

But the wonder wore off quite quickly. 'I then had got the feeling that doing computer programming was . . . a dangerous drug: it was something that you could get sucked into; and I spent an inordinate amount of time just playing. But it was great to have it there. However, the purpose for which it was bought, that never really worked.'

The reason for failure, after all that heroic effort, seems ridiculous. Brenner and White did overcome all the obvious obstacles. The printer worked as it should, and John White's automated worm-slice presenter, which brought slice after slice of minutely detailed worm under the microscope and photographed them, also worked. The program to collect and reassemble the slices would have worked, but the slices had been made by hand and each one was slightly distorted by the pressure of the scalpel. A human eye could spot this at once, and any computer you can buy nowadays could do it, too. But correction of that kind of visual distortion was beyond the capacity of the Modular 1, and White went back to another method. It turned out that human eyes were capable of very much finer discrimination.

In 1974, once the worm slices had been assembled into a coherent map, they were able to turn the computer into a kind of giant slide projector. By this time, White had built a digitising tablet and written the software to drive it, so it was possible to trace the sections into the computer and use it to project the pictures onto the wall. Fantastic stuff, for 1974. But the thought and judgment had all been human. It was another fifteen years before the computer programs surpassed humans in their judgment of the data, and by that time Brenner had lost interest in writing them.

However, the excursion into computers taught Brenner a great deal, and not just about the machines he was using. 'By thinking about computers, and by thinking about other complex objects, and about how one would try to account for them, I laid the ground for most of the way that I now view complex biological systems. I think increasingly everyone else will have to view them in this way . . . if you can't compute it, you can't understand it.'*

What he does not say is that he learned from real computing that nearly all his original ideas about biological computers were wrong, and that biological systems are impossible to compute in the sense that he first believed they ought to be. It turns out that there is no simple program for development which can be read off an organism's DNA. The Rosetta Stone that Bob Edgar had glimpsed was a mirage.

This may look like shocking apostasy. It is actually one of the most urgent lessons of the whole worm project. Most people who take an interest in science probably still believe, as Brenner did when he started out, that the DNA of an

---

* Brenner and Wolpert, *A Life in Science*, p. 62.

organism contains the program for its development, and that an organism, or even a single cell, is simply running a program in the genes. That was the original sense in which molecular biologists started talking about programs for the organism. The simplicity of the idea seemed to correspond with the deep and illuminating simplicity of genetic theory itself: that inheritance, like light and matter, comes in chunks; and when the word 'gene' was first coined it meant no more than 'a chunk of inheritance'.

The irreducible chunks of light and energy were quanta: hence quantum theory. The irreducible chunks of heredity were genes: hence most of the rest of modern science. And when genes themselves were broken down into chemicals, these, too, turned out to come in chunks. The only ones that mattered for the meaning of a gene were the four bases. It is a neatly digital system: a base can be only one of four things, and there is an irrefutable answer, yes or no, to the question of whether it is any one of them. Similarly, the genetic code that Brenner helped to work out comes with sharp-edged answers: each of the sixty-four possible triplet sequences of three bases can mean one of twenty amino acids, or it can mean 'stop'. But it cannot mean anything else. There is no twenty-second possibility.

The existence of a digital structure at the roots of life suggests that life is in some sense computation or the processing of information; that everything an organism can do is just the working out of a program stored in the DNA and that genes or their bases are simply bits of information in a humungous program. The bases strung out along a string of DNA look to be the same kind of thing as the patterns of holes strung out on the paper tapes with which the Modular 1 was programmed. If only we could read them right, it would be possible to see which holes said 'grow an egg here'; which ones said 'moult' or 'hatch' or 'mate' and how all these instructions, put into one long program, spelled out the whole life story of a worm; perhaps, if we could read them well enough, the whole life story of a human reader.

John Sulston, who came to be seen as the leader of the worm project when Brenner at last grew tired of it and went to study the puffer fish instead, sees the discovery of the digital nature of DNA as the one thing that gave biology a different direction from the 'harder' sciences in the last fifty years. As physicists have got more mystical, and seen their hard-edged atoms dissolve into clouds of probabilities, biologists have gone in the opposite direction and found clockwork at the heart of all the uncertainties of life.

'Biology is having its Victorian heyday of building up its clockwork. Certainty has been brought into biology precisely by getting down to that molecular level: that's what they did in the 1950s. I think that the amazing discovery that DNA is digital really epitomises the whole thing for me. That was an extraordinary thing. Who would have believed? People wrote whole books about how it couldn't possibly work: the coding of life had to be something much more holographic.'

Those books were wrong; and to Sulston that fact opens up a universe of possibilities. 'We don't know how far it's going to go. We don't know how much we're going to be able to actually compute an organism from its genes, although the fact that organisms do manage routinely to compute themselves from their genes makes one feel that we must be able to understand it. I mean, there's no point in people going round saying, "Oh, it's all uncertain and a bit of butterfly wings and Heisenberg." Crap! Organisms do it.'

The conclusion Brenner finally reached was that the simple program he had hoped to find could not exist in principle. It took him a long time to get there. Part of the argument hinges on the description of 'program' and the point is difficult to understand unless you have ever tried to write a computer program yourself.

What everyone who has tried to write one knows is that the essential thing about computer programs, no matter how small and simple, is that they don't work. Whenever you think you have reduced the world to unbreakable clockwork, a grain of sand appears and jams the mechanism. The literal-minded bloody dumb insolent stupidity of computers is not an accident. It is an expression of the inflexibility that defines a program's nature. There must be a unique, fixed, one-to-one correspondence between the input of an algorithm and its output.

Such rigid and deterministic control does exist in parts of the genetic mechanism. The genetic code itself is an example of an algorithmic program which really could be mimicked in a computer. In fact, there are undergraduate courses which have students do exactly that, using carefully shaped bits of paper which fit together like jigsaw-puzzle pieces to represent the specific way the different molecules of DNA and RNA recognise each other. Any given codon or triplet of bases will produce either exactly one amino acid or else an instruction to stop. The rigid and programmatic nature of the code explains why there are so many mutations in nature. Changing even one input will almost always change the output.*

Development is more flexible than that. It is true that there are islands of order, or subroutines, in the way that worms grow and very large developmental programs can be

---

* An exception would be a codon which differs by one base from another which codes for the same amino acid: e.g., AAA and AAG both code for phenylalanine.

switched on by very few genes. Sex, or gender, is an excellent example, because it makes huge changes in the way a worm grows, yet these appear to follow in a programmed way from rather small decisions: about 30 per cent of the body cells of a male worm are different from those of a hermaphrodite, yet only about twelve of the worm's 19,000 genes are involved in deciding which sex an egg will grow up to have. A possibly more familiar example is 'hox' genes, which determine the patterning of embryos, from front to back, in almost every organism that has a front or back, from worms and flies to humans. Some of the ways in which worms grow look very much like the repeated operations of subroutines, in that the patterns of cells grow repeatedly down the body. An even odder and more suggestive fact is provided by the way cells die within these subroutines: one in eight cells produced as the worm grows from an egg will die a programmed death. Some have functions fulfilled before dying, but many do not. The oddest fact is that almost all programmed cell deaths are suicides, in the sense that they can't be prevented by burning away the neighbouring cells with a laser.* The cell appears determined to die from the moment it is born. Why? One theory is that these programmed deaths show us the boundaries of the subroutines. In a real computer program, it is trivial to say, 'Copy this five times.' For an animal whose cells split in two, it may be easier to make eight copies and throw three away.

The subroutines are not themselves determined by a genetic program. Sometimes they are switched on in genes with a particular ancestry, sometimes they are a response to the circumstances in which a cell finds itself. But once such

---

* There are two deaths which are known to be murders, because they can be prevented by the prior destruction of other cells.

a subroutine has started, the cell will go through it to the end.

Once you turn from development to maintenance, the picture grows more complicated. The number of genes involved in making an animal grow, and choosing its shape, is really quite small compared to the number involved in the processes required to keep it alive, and these are not susceptible to any central program. Even development is flexible when you look at it closely.

Although all living things are made from parts specified rigidly by the DNA jigsaw pieces, their defining characteristic, as they grow, is flexibility. A properly developed worm has 302 neurones, yet a worm with only two of them can live; similarly, a worm can compensate for the loss of almost any of its muscles. This kind of flexibility seems quite beyond the capacity of a single deterministic program to manage, even if it can be produced by the interactions of numerous small, discrete, deterministic programs. Biological computing is massively parallel, and analogue, whereas the computers we use are digital, and serial.

The distinction is brought out most clearly by looking at the kind of program that would be needed to run an organism which is also a single cell and therefore about as simple as anything interestingly alive could be: *E. coli*, the bacterium that worms eat. You might look at one of these bacteria, Brenner argues,* and see it as a small chemical factory, which must copy itself once every thirty minutes. Within the factory there are several thousand chemical reactions going on simultaneously, all mixed up with one

---

* The clearest formulation of this argument can be found in his talk published in the seminar volume *The Limits of Reductionism in Biology* (London: Wiley, 1998).

another; almost all are necessary if the bacterium is to build a new copy of itself.

If a human were to try and program such a factory, there would need to be a controller which could keep track of all these simultaneous reactions; which could ensure that each one had the materials it needed, and that the right quantities of the right chemicals were being produced at all times. Of course, in a real factory there would have to be pipes carrying all the different reagents around and ensuring that the output of one reaction got piped through to the next in time, that the speed of delivery was just right, and that the different temperatures needed were precisely maintained.

The point is that this cannot be done. It is impossible to understand all that is going on, and all the complex inter-relationships, well enough to build a single controlling program which could run the factory reproducing the bacterium's activity. Each reaction might affect many other reactions. The task of monitoring all of them, and then working out what the system should do next, and checking that it had actually done what it should, before deciding what the next step again should be: all that is beyond the capacity of even the most powerful computers imaginable. 'The main problems are the enormous amount of computation required to maintain the system and the cumulation of inaccuracies. Even if the central machine is assisted by lower level computers that will translate the sensors or control the valves, it will have to supervise them as well as administer itself and this alone consumes a large sector of the computational resource . . . It is known that when such systems reach some level of complexity, it becomes difficult to maintain them in a steady state: they become metastable, and either run away and blow up, or lie down and die.'*

* *The Limits of Reductionism in Biology*, p. 106

You might think a controller program is needed to maintain order among all the contending chaos of the cell, but in fact, says Brenner, the opposite is true: 'Top-down mechanisms impose incoherence because you can't repair them.'

Yet building these vast controlling programs is the traditional route of artificial intelligence. It is certainly how most people imagine the program in the genome. Whatever their limitations, computers can control and keep track of systems vastly more complex than the human mind can manage. That is why telephone exchanges are now full of computers and not switchboard operators. But when John White turned down a job designing telephone exchanges for the worm, he was moving into a different order of complexity, which artificial intelligence cannot match.

So how does the bacterium do it? And how does the vastly more complicated worm? The answer, Brenner says, comes from the way biological chemistry works. A human factory has all its reactions going on in separate, defined places. Inside a cell, by contrast, everything is jumbled up and simultaneous. What separates the differing reactions is the fact that proteins have shapes and their complicated bumps and wiggles means that each one will fit only into certain others. Proteins in their individual gnarliness actually are the way that people in love believe themselves to be: able to fit happily with only one, rare, right partner. So all these monogamous proteins can wander around and they will find their perfect partners; whereas if you were to design a factory it would be much more like breeding dogs. You would have to spend more energy on preventing unwanted reactions than on ensuring that the right ones happened.

To put it in computing terms, proteins in a cell can have logical addresses rather than physical ones. Because every molecule wandering around inside a cell will bump into every other one, yet fit together with very few, it does not need to know where to go. Its fate or function will find it wherever in the cell it may be.

There is one more factor which makes biological computing manageable: the reactions in a cell naturally arrange themselves in feedback loops. The final product of a long chain of reactions (or computations, if you want to think of them that way) will have an inhibiting effect on the first link in the chain, so, once 'enough' of a substance has been made, the reactions that made it will stop. How does the cell know how much is enough? Natural selection has instructed it. Every possible answer to that question is contained in the properties of one protein or another. Some stop the reaction earlier, and some stop it later. Which one gets made in a cell today depends on which of the innumerable trials in the past was most successful. In that way, the DNA stores huge amounts of answers to questions asked in the past, rather than any mechanism for finding general answers. The program the cell carries out is very simple: it is just to look up the answer to any question in the right bit of DNA.

So what is left of the dream of finding and copying a worm's developmental program? The answer, Brenner still believes, is that the end of the worm project should be a computer program but one quite different from the one he first thought of. What is needed is not artificial intelligence but 'artificial stupidity': using human intelligence to model the autonomously functioning parts of the worm in all their limited particularity, and then watching how the models interact. This plan goes back to the very beginnings of the idea of a biological program in the work of Johnny von Neumann.

Von Neumann's original idea was that if there were things too complex to talk about in language, it would be easier to model them in computer programs and see how close you could get to understanding that way. You discover the flaws in your model when it fails to react as the original does. But this is always a concrete discovery, one which defines the problem to be solved. It is not just a 'mystical' lack of knowledge. Brenner does not like mysticism.

Complete understanding of the worm, he says, would mean being able to write a program which would model all the ways in which a worm can react to anything, and would do so because it contained all the properties of all the parts of a worm, and all their possible interactions. It would be written in the same 'machine language' as the worm. You would start with bits of software which modelled the behaviours and properties of proteins, and then see how these might fit together little subroutines, like the activities inside a cell, where each chain of reactions is controlled by a natural feedback process, in that the chemical produced at the last stage slows or inhibits the first stage.

At last, and with immense effort, you might have a simulation of the processes inside a single nerve cell. There are three hundred different nerve cells in the worm, all of which would need individual modelling; and the modelling would have to take into account the history and fate of each nerve cell. Then there are the muscle cells, the cells of the skin, and the immense complexities of the gonad, and the fertilised eggs inside the worm, each of which contains, in some sense, the programs needed to make a new worm.

Brenner is sure that in principle it could be done. 'The real thing I'm against now is this idea of emergent phenomena. It seems to me very mystical. They say that if you mix things that get to a certain level of complexity, then you get emergent properties. That's to try to say the whole is more than the parts. But, in fact, the whole *is* the sum of the parts and their interactions, that's all; and our job is to find the interactions. Then we could compute the behaviour of the system.'

It is not at all clear to me whether the people who do believe in emergence would disagree with Brenner here. Like other large terms in the philosophy of science, such as 'reductionism', it is much more use in practice as a shibboleth than as a practical definition. We may not know quite what it means, but we can be sure we are against anyone who believes in it. Such low impulses give people the strength to push their arguments to useful conclusions, and this argument goes to the heart of Brenner's idea of what science is for. In everyday English, we may say that the results of a calculation 'emerge', but to scientists of Brenner's temper this usage is a frightful discord. 'Emergence' to him entails something which is impossible in principle to calculate and not merely difficult to calculate in practice.

A quick trip to Discworld may make this plain. In our own universe, I can say that a picture emerges when I develop the camera film. But this is not emergence in the obnoxious sense, because anyone who knew the chemical constituents of a spot of emulsion on the exposed negative could calculate exactly what shade that spot would produce when developed. In Terry Pratchett's universe, however, the picture emerges because in the camera there is an imp who paints it. That is incalculable. Those imps are the creatures Brenner thinks science should banish. He thinks science is for predicting the behaviour of the world by discovering its underlying calculable regularities. 'Once I have a model, I can test its validity by changing it in some way and then look for the predicted outcome by making the same change in the real system.'*

* *The Limits of Reductionism in Biology*, p. 106

'The programme is to be able to calculate the future,' he says. 'That's what science knows. Otherwise we're just scribes.

'We want to be able to predict behaviour under circumstances because we cannot go and measure behaviour under all possible circumstances, I mean, we used to say, "Look, you see, we could measure what happens if we put the worm in hot water, cold water, slippery surface, etc., etc. But if someone says, "What happens if I put it on a satellite circulating the moon?" what I'd like to be able to do is to say, "I'll tell you what will happen. I'll calculate that. I know the inputs." Unless we are going to have that as an objective, then it's not science. People need to be reminded of that.'

Looking back on the worm programme long afterwards, Brenner said that he wanted only people who would take risks with their career: it might all have fizzled out into nothing. But that is not how his first followers remember it. For them, the risk was negligible. Brenner's project was just too exciting to pass by, and the only problem, it seemed as they fell under his spell, was that they might not have enough time to do all that needed doing, or to discover everything concealed in the tiny lives of worms.

4

The Nerves

Most of the work on the worm was done by people of exceptional energy and drive. None of them – except perhaps Brenner – had had a career which ascended in a straight line, if only because the worm, when it started, was off to the side of any imaginable career path; but all the crucial early workers ended up at the summit of their professions. John White is at the University of Madison, Wisconsin; John Sulston ran a third of the public human genome project and Bob Waterston ran another third; Bob Horvitz is at MIT. Brenner, Horvitz and Sulston shared a Nobel Prize in 2002 for their worm work. Eileen Southgate, however, lives in a small house outside Cambridge, as she has done for many years. She went to the lab as a technician, and retired as one. Yet her contribution to one of the great worm projects was as immense as anyone's. She did get credit for it, both in conversation and in history; this is not a story of a woman overlooked by the scientific establishment. But it is very easy to overlook the role that is played in scientific advances by a disciplined lack of imagination and a simple willingness to trudge onwards, doing what amounts to high-class factory work.

Eileen Southgate had worked at the LMB since leaving school at sixteen in 1956, when she was offered the chance to help, as the careers officer said, with medical research. 'It was a choice of three jobs from school, actually. It was in the days when they came to the school with jobs and said, you know, you have a choice of three. One I couldn't get to, the other one I didn't really want, and I chose this one. There was another girl there I'd been to school with, whom I knew, so . . .'

What did she like about it? 'It was easygoing. It was then only, of course, a small group of about thirty people so it was very friendly, just like a family, really.'

She worked for Max Perutz and John Kendrew for much of the patient labour on haemoglobin and myoglobin that won them Nobel prizes: there are champagne bottles signed by all the prize winners from the LMB on the shelves in one corner of her living room, though – and this is a touch that tells almost everything you need to know about the LMB – she can't remember which prize party they came from. But the excitement of their work did not transmit itself to her: she made up the chemical solutions and ensured the stock room was under control. 'Every year it changed a bit,' she says, but it's obvious that it didn't change a great deal. Then, in 1969, she was assigned to work with John White, himself a former technician, who had been given, as the second half of his PhD project, an examination of the nervous system of the worm.*

Brenner's quest for the links between genes and behaviour concentrated on mutants that interfered with movement, which is about all the behaviour that a worm can have. By 1974 he would find more than a hundred *unc*

---

* The first half was the computer programming with Sydney Brenner.

mutants, as all those were called which disrupted the fluid sinuous writhing of a healthy worm (*unc* stands for un-coordinated). All of them could disrupt entirely different parts of the nervous system; some might simply disrupt the formation of muscles. Either way, if they wanted to under-stand in detail how behavioural mutations changed the worm's anatomy, they needed a map of the worm at very fine scale.

They didn't mean at first to do the whole of the worm. Originally, Brenner believed it would be enough to examine one small area of the nervous system in great detail, because he believed that any mutational changes would be bound to show up there and that you could take for granted similar changes elsewhere. The part he started with, the retro-vesic-ular ganglion, is one of the more complex bits of the worm's nervous system, with about sixty neurones clus-tered together, but it did not look impossible to make into a model* in the way that the whole worm must at first have seemed. 'I had no idea when we started what we were get-ting into,' said Nichol Thomson. Around the pharynx there is a sort of wreath, or knot, of nerve cells, which is some-times called the brain; two long strings run back down the animal from there, one ventral and one dorsal. There are sensory nerves running forward to the bumps round the mouth, and the male has an extra, complex extension of the nervous system back into the tail to help it mate. But most of the nerves are in the knot and the two strings, and almost all the complexity is in the knot. The retro-vesicular gan-glion comes just behind the ventral ganglion where the ventral cord runs into the nerve ring.

They mapped enough of the retro-vesicular ganglion to realise that the work was technically feasible. The worm

---

* Jonathan Hodgkin, interview with the author.

could be sliced thinly enough, and these slices photographed under an electron microscope in ways that
allowed a sort of three-dimensional jigsaw of the ganglion
to be assembled. At one stage they made a plywood model
of part of the nervous system, Southgate remembers: 'We
had to trace all the sections, then trace them onto some sort
of plywood. It was very weighty. We only did small sections – I think the biggest model was about two feet long –
but then I painted all the different cells different colours
and when it was all put together, well, it looked rather
pretty, actually.'

After a while the ambition of the project grew immensely.
It had to, by the logic of Brenner's grand plan: he wanted to
show how the traffic of signals though the nervous system
gave rise to behaviour, and to do this he had to be sure he
knew exactly what the nervous system was. He had
invented a figure he called 'the sceptic', who stood at his
shoulder and, every time he claimed he knew how some
signal was carried around the worm, would ask, 'But how
do you know there is no other wire involved?'

'I came to realise that what I had to do was to be able to
quench the sceptic.' He needed to be able to reply to this
hectoring onlooker that there could be no more wires
because he knew the location of every nerve and all its junc-
tions in the whole body. That would mean modelling and
examining the whole hermaphrodite, from tip to tail, at a
level of detail which meant it would take years. No one
realised at first quite how many years would be involved.

Finding and counting the wires was the task that Eileen
Southgate found herself working on with John White and
Nichol Thomson. She should have been redundant from
the start. If the great machine John White tried to build had
worked, the slices would have been photographed automatically, and the computer would have lined up each

photograph so that it was obvious how the connections were made between slices. But in practice things did not work out. Her eyes proved sharper than anything a computer could manage, and she settled into the routine of preparing and photographing the slices Thomson had made ready.

He worked with worms which had been preserved chemically, dried out, and then embedded in blocks of araldite. Working under an ordinary microscope, Thomson shaved the araldite block as closely as possible round the worm, ending up with a tiny roll of araldite, almost too small to be seen with the naked eye, and with a single worm embedded in the middle. Then followed the really fiddly work. The araldite block was mounted in a microtome, a kind of miniaturised sawmill, which held it in front of a diamond-tipped knife beneath a microscope. Normally, in this kind of work, glass knives were used, but they wore out too quickly for the kind of shaving that Thomson was doing. What he was doing was not dissimilar, in principle, to the way that logs are sawn in factories, except that the scale was absolutely minute.

Thomson was slicing the worms into salami one twentieth of a micron thick. If you have nicely trimmed fingernails, there is a white brim at the front about a millimetre thick.* That's about as long as a worm, and Thomson was slicing them so slowly and finely that it would have taken him 20,000 slices to trim the white line off the end of a fingernail. To get the worm to move in such fine increments, you could not adjust it by any normal mechanism like a screw. Instead, an iron bar was heated, and as it expanded, quite undetectable to the naked eye, it

---

* In children this white bit is actually black. Adjust the illustration accordingly.

pushed the worm forward until another twentieth of a micron was ready for the diamond knife. This mechanism imposed its own rhythm on the work, because when the bar was fully expanded it had to be switched off to cool down, and the worm readjusted, very, very precisely. 'If you missed half a micron you would have to throw the worm away.'

For the continuous runs of really difficult work, Thomson learned to get 3,000 consecutive sections out of a single worm – his record was six thousand – which took him about three days; one good sneeze could have destroyed a week's work. Then the shavings had each to be mounted in tiny wire grids, in order, and stained with chemicals which made their different parts stand out clearly under the electron microscope. 'Of course, each time they were stained with something, you had to wash them and dry them and it was delicate work. However, I seemed to be able to do it so they kept me on.'

These were the grids that Eileen Southgate had to place under an electron microscope and photograph before making tracings which showed where everything of interest was. She had done some slightly similar work in preparation, years before, looking at slices of *Ascaris suum*, the large relative of *C. elegans* that had been the first nematode whose nervous system it made sense to study. But those had been much thicker slices, made with a normal microtome, and they were examined under an ordinary microscope.

The first thing she had to learn in her new job was how to use an electron microscope. She loved theirs because 'It was so old it was really simple.' It worked with boxes of twenty-four glass plates at a time, arranged like a rolodex so they were exposed one after the other. She cranked the handle and a picture was taken. For particularly fiddly or complicated bits of the worm, they used four plates across and printed them all out as big as possible – three feet square, if necessary – so that Brenner's invisible sceptic would have no cracks to hide in.

There was a plan to do that part on computers, too. 'We had an American who had the bright idea of putting everything onto the computer, but to print it out you had to have sheets of paper the length of this room, whereas I could get it all on an A3 graph paper. It was ridiculous. John and I tried one that way and then gave up.'

So it was back to the steady grind of printing the photographs, tracing out the structures they showed, and making sure that every single connection between the neurons was numbered and identified. On a good day, she could do a hundred photographs like that, or five-thousandths of a millimetre of worm. The pictures that resulted from all this have a unearthly and irregular beauty. The interior of the worm becomes a pale blue world, with darker threads running through it. When you study them, they have the quality peculiar to scientific photographs that without the key they could be anything at any scale: nebulae can be photographed so they look just like the interior of nematodes. Distinguishing the features they were looking for is hard for an untrained eye. A chemical synapse looks different from an electrical one; a fibre which leaves one neuron 5,000 slices into the worm might reach its destination only 200 slices further in, and must be tracked, among all the other fibres, through all the intervening photographs.

Southgate loved doing real jigsaw puzzles, too: people in the lab used to joke that she spent all day fitting together the worm slices to discover the nervous system and then went home and relaxed by assembling pictures of Salisbury cathedral. Other people helped from time to time. Brenner himself had started the work and did some sectioning. Jonathan Hodgkin lent a hand in the early days but found it unspeakably boring. Still, in the end Southgate and John

White sat in one small room at the MRC and studied every one of the 20,000 sections of the worm by eye, then drew on paper diagrams of every single nerve and its connections. There's no doubt that she did most of the really monotonous stuff. John Sulston, who worked down the corridor, said, 'John actually could do her job and did, in fact, if he was sufficiently motivated. And this is important. The Chief must always be able to do the Indian's job and appreciate what goes into it. That makes for a good management structure, doesn't it?'

Sulston really did see Southgate and White as partners because both of them were in thrall to the puzzles they worked with. This may be the deep underlying link between the satisfactions of science and of computer programming. In both cases, you are pushing intangible pieces of the world around until suddenly they click into a deeper and more satisfying pattern. 'People lie awake at night,' says Sulston, 'fitting these pieces together, and then they go in the lab and they actually do see that one of their thoughts came right. The pieces go click and it's tremendous.'

In this sense Nichol Thomson, the universally admired virtuoso, was a less typical scientist, perhaps, than Eileen Southgate, wry, shrewd, phlegmatic, grinding away at a job which most people in the lab felt they couldn't do because it was too monotonous and perhaps beneath them, work which even she described as 'repetitive and boring.' But she was constantly solving small practical and intellectual puzzles, understanding a little better how the world is fitted together, and making this knowledge available. Not that she thought understanding the worm was a particularly sensible thing to do. Did you ever think, I asked her, that you knew more than anyone else in the world on this subject? Yes, she said. But she wasn't very impressed: 'I never really thought about it that much. No, they used to say, "You're the only person in the world doing this," but I used to say, "Only one mad enough!" It didn't seem . . . I mean, it was a job; I was paid well for doing it.'

It's true, though, that no one will ever know as much as she did about the nervous system, and no one will ever check the work she did. 'We had several Americans came and they were full of . . . Oh, they wanted to find out this and that; and then, when they realised what was involved, they changed tack a bit. They did little short bits. That was all.'

Long before the map was completed, it was useful. It was never meant to be a piece of self-sufficient knowledge, but was intended to reinforce the other ways of understanding the worm. This mutual dependence was apparent from the very beginning, in the way that the physical studies of the nervous system were meant to complement the logical dissection that the genetic program brought about. Once John Sulston started watching the particulars of the worm's growth patterns, and tracing the lineage of each cell all the way back to an egg, this became the third strand in what was by now a braided understanding of the worm, much stronger than any of its individual parts. The physical map of the worm, and of its growth, was complemented by the logical and functional maps produced by genetics, and the historical maps produced in the lineage charts. Of course, this did not mean that the worm was well understood: twenty years after the map of the nervous system was finished I attended an international worm conference in Los Angeles at which nearly 2,000 papers were presented, all adding little discrete chunks to the world's knowledge of how things are done in a worm at the molecular level. But the areas of ignorance are very much smaller now. If you go to study the worm, it is with a clear idea of what is not known about it.

It may be that the mid-Seventies were the time when the worm, for itself, was most fun, because just enough was known to make it clear where the next discoveries might be made. The long, grinding work of the first years was over. Brenner himself withdrew a little. He no longer gave his performances every day in the coffee room, which disappointed some of the newcomers. On the other hand, he no longer needed to sparkle and glitter to draw people into his field. The work itself would do that. Once the maps of the nervous system provided a scaffolding, it was easy to climb a little way up and see what needed doing next. For a certain sort of temperament this was simply irresistible.

Judith Kimble is a woman who quivers like a candle flame in my memory: a warm uprush of enthusiasm and illumination. She must have slowed down a little now, in her early fifties, because you can actually see her at rest: she is a professor at the University of Madison in Wisconsin, one of the major centres of worm research. Originally, she planned to be a medical doctor. Well, first she wanted to be a veterinary surgeon 'because I liked cats and dogs. I couldn't pronounce all those syllables when I first decided I'm going to be a vet, but when I was four, before I could say "veterinarian", I learned that "vet" would work.'

As a teenager, she became more interested in human beings. 'I was so anxious to become a doctor that I was taking all the courses that you do at medical school early. I was taking anatomy and physiology and some biology, even though I didn't want to take all this biology stuff because I wanted to be a doctor.' It was actually an anatomy lecture at Berkeley in 1970 that turned her into a biologist. 'I'll never forget: one of my anatomy professors who was teaching development talked about how you could take heart tubes when they were just the tubes and put them in a petri dish

and they would bend the right way and form a heart.' By
that time she had applied to medical school and been
accepted, but she could not get the story out of her mind.
She drove clean across the continent in eighteen hours,
'thinking about whether or not I really wanted to go to
medical school or wanted to understand something as unbe-
lievably amazing as how a heart tube could form a heart in
a petri dish. And I decided that I really wanted to do the
latter.'

She was too late to get into an American graduate school
in biology, so she went instead to Copenhagen, where she
had a lot of fun and studied vertebrate embryos under electron microscopes, but without, she felt, getting any nearer
to the mystery of what made fragments of tissue grow
together into the shape they would need as adult organs.
What she was taught in Copenhagen was entirely descrip-
tive. It told her what happened, but without giving any
handles into how or why. So she started to read ancient
textbooks of embryology, to look for clues to the mystery.
She decided the answer must lie at the molecular level but
as far as she was concerned molecular genetics was still
something that was studied in the smallest possible pack-
ages: *E. coli* or the phages that prey on it. She had never
even heard of the worm.

After Copenhagen she found a place in graduate school
in Boulder, Colorado, working with David Hirsh. He had
caught the worm off Sydney Brenner in 1971, when he had
been a post-doc at the LMB, working on tRNAs. It was just
that the worm seemed so much larger and when he went
back to the states, as an assistant professor at Boulder, he
started a worm lab, the first outside Cambridge (the spread
of the worm is the subject of a later chapter). For Judith
Kimble, the Boulder lab turned out to be a way into LMB,
where she arrived in 1977, moving into John White's house

just as he went off to spend a sabbatical year in Boulder, and she found herself completely happy.

'It was the best time of my life, I think. I mean, I wasn't able to sleep for the first six months – literally. I was just so excited to be there that I would go to sleep at three or four o'clock in the morning and I would be up at eight – and after about six months of that, I realised this was not smart.' She found everything about the lab wonderful: not just the camaraderie among the researchers, but the freedom from bureaucracy. 'Nobody was drained away by all of the stuff that we get drained away in university: administration, teaching, committees, all that stuff. I mean, it gets to you, writing grants. But if they wanted a microscope, they put it all on a piece of paper and, apparently, just gave it to Michael Fuller downstairs and got it.'

I often wonder, when people tell me about those days, how on earth anyone found the time to have babies. They certainly did: Sulston was raising two children; Phil Anderson, another post-doc, had a son in Cambridge; John White and Donna Albertson had two children together. But the strain must have been immense. Judith Kimble was not exaggerating the hours people worked, or at least the hours they spent in the labs: the evidence is there in, for example, the logbooks kept by people watching cell division in the living worm. By her own account, the day started at nine in the morning: she puttered around until 10.30, then there was coffee and conversation for three-quarters of an hour. Then another couple of hours in the lab, then lunch, and more coffee: after about an hour and a half, maybe two hours, of talking to people just as smart and dedicated and absorbed in their work as she was, she went back to the lab. Then there was tea. The real work, she says, did not start until around four in the afternoon, when people settled into their little bench spaces and actually ran the experiments they had been talking about. By ten in the evening, everyone was exhausted and hungry, so they crossed the road to the Addenbrooke's Hospital canteen, where there was a fuelling meal available for 19p ('I'm sure you can imagine what that tasted like!'). That kept them all going until midnight, when she went home. And she kept it up, seven days a week, for six months. 'Maybe on Sunday you would do the laundry.'

That kind of exhausting, exhilarating and completely satisfying grind seems to have been the normal condition of almost everyone who worked on the project. It does throw into very high relief the degree to which the project depended on people like Eileen Southgate, whose hours were not nearly as grotesque, but who did her work, year in and year out, without the sustaining champagne-in-the-veins delight that kept the scientists going. Muriel Wigby liked most of the scientists on the worm project, but from her perspective as a lowly geneticist they were not good breeding material. 'Most of the scientists were not good husbands, and very bad parents. Apart from being rich and fertile, they did nothing for their wives. Some people never do anything but science. They just won't invest enough in their relationships.'

White never stopped tinkering. By the time he went off to Boulder he had managed to hook up a laser to a microscope to produce a kind of laser scalpel that let him pick out a single cell anywhere in the worm beneath the microscope and burn it away with complete accuracy. The work was not without its casualties: one lunchtime Bob Horvitz, an American post-doc, came back to the lab to find the lens of his expensive microscope no longer transparent after White had managed to focus the laser on the optics rather than through them. But, once it worked, this precision scalpel became one of the most successful and versatile tools in the worm unit. Apart from anything else, it was the closest that technology offered to an arcade game. Judith Kimble remembers that it was almost the same procedure as an ordinary microscope: she used to place a worm on a tiny patch of bacteria on a slide, as if she were going to watch it growing, focus on the cell she wanted to kill, and zap! 'I got to a point where I'd get them on the run because I was too lazy to anaesthetise them. You know, like shooting practice: it was really fun.'

The wonderful simplicity of the technology – at least at the point where she would shoot – contrasts with the enormous complexities of the process she was discovering with it. One of the things that interested her was where the vulva comes from: what cells must exist in the earliest larvae if they are to grow a proper vulva in their later stages. One obvious candidate is known as the anchor cell: it sits in the gonad, and the vulva develops right underneath it. Sure enough, she shot the anchor cell out of an L1 (first-stage) larva with the laser and its vulva failed to develop. But where did the anchor cell itself come from?

Here the picture gets absurdly complicated, because the formation of the anchor cell is one of the very few moments in a worm's development when no one can predict what exactly what will happen next. There are only two cells that may become an anchor cell; but only one of them does, in a normal worm. If you destroy either with a laser, the other will always become an anchor cell. If both survive, they decide between them which will do what job, by an exchange of signalling protein; and which cell ends up sending the signal, and which transmitting it, seems to be wholly random. Both of them start with two interesting genes active, *lin-12* and *lag-2*. *Lin-12* encodes a protein which passes right through the cell wall, like a stud: at its outer end there is a macaroni whirl, or binding site, which fits exactly into a part of *lag-2*'s protein. Both cells start by producing both kinds of protein, but some random event always tips the balance, and each one ends up specialising either in *lin-12* or in *lag-2*. Part of the mechanism appears to be that the *lin-12* protein not only fits *lag-2* outside the cell, but, when it happens, also changes the shape of the bit of the *lin-12* protein inside the cell in ways which make it stimulate the production of itself, and shut down that of *lag-2*. It's a nice example of the ways in which the three-dimensionality of proteins means that they can fit – literally – into lots of different chains of cause and effect, and so carry different messages around the cell.

The cell in which *lin-12* is produced goes off to become the ventral uterine precursor: the cell which emits *lag-2* becomes the anchor cell, whose position determines where other cells will start to grow into a vulva. Here again there is an element of flexibility, though not of chance. There are six particular skin cells which may start to divide until their descendants have formed themselves into a vulva: any cell with that particular ancestry is fitted for a vulval destiny. But which ones actually assume the responsibility is a matter of negotiations among coevals. Three of the possible candidates have to go off and do other things in the body: if they do not, you get the multi-vulva worm, which has three in a row down its belly. The genes implicated in this catastrophe have names that start with MUV. If there is one thing a worm researcher cannot resist when a new gene is found, it is a slightly dirty joke: when a gene was announced which caused the male to have difficulty finding the hermaphrodite's vulva, it was named LOV for Loss of Vulva. But this is jumping ahead of the story: the point is that, as she started to play with the laser, Kimble could see that there were immensely complicated and wonderful things going on which would take years to work out, and it was all going to be tremendous fun.

There were some very strange discoveries made in this period, too. Sulston and Jonathan Hodgkin found – and knocked out with the laser – a tiny little cell beside the excretory canal, which didn't seem to do anything they could understand. Tendrils extended from it towards the head and tail, but no one knew what they did, either. It shouldn't have been important, since worms can grow to maturity without an excretory canal at all. But when they killed the cell in newly hatched larvae, the animals shrank at the back, appeared half-starved and died shortly afterwards. They never did find out what this cell did, except that it was vital for life, so they suggested they'd found the nematode's soul.*

Many years later, after the worm's nervous system had been exhaustively mapped in the old way, White at last managed to build the four-dimensional microscope that was more or less what Brenner had wanted twenty years before. Using this device, which was hooked up to a sophisticated computer storage system, you could watch and record what was happening within a worm at different depths, and then play back the films in any way you wanted. But it worked with microscopes which used light and so had a relatively coarse resolution. Nothing has been made that uses the electron microscopes that White and Southgate needed for their vast project.

It took them fourteen years. In all that time, she never worked with live worms. Once all the connections were known, and all the photographs taken, she traced over the photographs using plastic marked with grid lines to come up with her final – really beautiful – schematic drawings.

* *Worm Breeders' Gazette*, 5:1.

That took another four years, as she remembers. Then there was a pause of two or three years before they could find a journal with enough room for all they had discovered. It finally appeared as a special, hardbound issue of *The Philosophical Transactions of the Royal Society*, 340 pages long. The working title was *The Mind of a Worm*, though other people called it 'The Most Expensive Wiring Diagram in History'; either name is recognised. There was no party to celebrate the publication.

John White got a professorship in Madison, and moved there in 1983: the plywood models of the retro-vesicular ganglion were somewhere in the 128 crates that Eileen Southgate had shipped out to him after he had gone. She took early retirement; after twenty-one years of the worm she didn't want to work with anyone else.

It is almost certain that the task will never be completed with the male anatomy. Phil Anderson, one of White's colleagues at Madison, says: 'Very few people could, even today, do for one synapse or one cell John White did for the whole animal, it's just too labour-intensive. The methodologies are just too cumbersome, too expensive, and few people in the world can actually do it.'

5

Sulston and the Cells

If Sydney Brenner in full flow reminds one of an alpine river, bounding along in an invigorating froth of brilliance, John Sulston has the self-sufficient buoyancy of a dry fly, able to ride the craziest rapids. That is not just because he is well hackled, with a badger beard and hair which – though now trimmed – used once to form a halo round his face. There is also an unsinkable sturdiness to his manner; a kind of enthusiastic competence which inspires trust.

He did not at first seem destined for any sort of leadership. He came from the kind of high-minded middle-class background nowadays preserved largely among Quakers. His father was an army chaplain (an Anglican) who became the secretary of a rather low-Church missionary society. His mother was a teacher of English. Like John White, he was a fiddler and a tinkerer as a child: a passion for clockwork seems to distinguish molecular biologists. More than any other of the people I talked to, Sulston is unashamed about his pleasure in toys, in mechanisms, and what he calls that 'artisanal' side of a scientist's work. He was the first man to learn the trick of freezing worms so they could be thawed out alive. Actually, this works best with very small larvae, but in any case it was an absolutely necessary technical breakthrough if the worm was to become a reference animal for biology. Without the frozen reference stocks going back to the 1970s the worm could have been quietly evolving in laboratories or accumulating all kinds of deleterious mutations; but if that seems to be happening now, it is simple to wash away all the suspect worms and thaw out a fresh batch of the original vintage. Equally, the ability to freeze mutants of interest means that they are always available as soon as anyone needs to investigate one.

As an adolescent, Sulston lost the religious faith of his parents but he maintained their ethical seriousness combined with an anarchic streak. He became a scientist almost by accident. He had worked hard for a 2:1 degree in chemistry at Cambridge, but decided when it arrived that 'book learning wasn't for me' and prepared himself for a stint as an aid worker with VSO instead; that fell through at the last moment. It was the early Sixties and universities were growing fast. He picked up a PhD project in Cambridge 'just by wandering along the corridors'. He discovered that he loved it and though he never planned a career in science, neither did he ever find anything he liked doing half as much.

By the time he got his doctorate, he had found his other binding site in life: Daphne Bate, a research assistant in another department. They got married when he was offered a job in California with Leslie Orgel, a chemist who had just moved there from the MRC at Cambridge, where he was one of the inner group: he had been one of Brenner's travelling companions in spring 1953 when he drove across from Oxford to Cambridge in an unheated car to meet Crick and Watson and see their physical model of DNA. Sulston's job in California was to investigate the chemical beginnings of life, the very first molecules that could reliably be copied in the primal soup. In Orgel's lab, at the Salk Institute in La Jolla, he was moving among the elite of science. Francis Crick used to drop by to chat, and Sydney Brenner, too, though these chats never felt like job interviews.

'One felt spoilt there by all the people coming through. After a while, you get used to it and you take it for granted – in a reasonable way, I think. It's not that you feel you're part of the top echelons but simply that you don't make so much of a fuss about it.'

The Sulstons had a daughter, and partly as a result lived a quietly bohemian life: what with the hair, the beard and the jesus sandals that he wore for years, he is often typecast as a hippie, but he wasn't the urban flash sort: more the earnest rural Whole Earth type, growing vegetables in the back garden partly because it is economical. They loved America, driving all round the continent in battered old cars, but after three years Daphne wanted to go home. He could have had a post in California, but he also learned that some of the conversations in the lab had in fact been job interviews; so he was offered a job with Sydney Brenner at the LMB. He was hired as a chemist, with various specifically chemical tasks in mind. One of his first jobs was to estimate the amount of DNA in the worm by purely chemical means; a measurement which turned out to be remarkably accurate thirty years later when it was all counted digitally.

He made little impact on the lab when he first arrived. 'There was that beard!' says Muriel Wigby. 'In those days, he was a very quiet man. Unless you actually spoke to him, he seemed to be very shy, hiding behind all that hair. But when you talked, you discovered that he had a lot to contribute.'

Like everyone else, Sulston found the full flow of Brenner awe-inspiring. 'Sydney does very much go in, and always went in, for the monologue: we'd all be sitting around after lunch with coffee or whatever and he would hold forth quite extensively; others would say a bit but it was mostly Sydney talking. He will also constantly throw out ideas, far more than anybody can cope with. It's a jumble, you know. He does a lot of thinking and originating of ideas for himself; he also is a great passer-on of gossip and it's impossible to distinguish the two.'

The effect of this, however, was wholly beneficent: Brenner 'really strongly disliked the idea of being bound by rules' but underneath the glittering brilliance there was a strong current of real scientific communication. Conventional labs have group seminars, where everyone makes little presentations about their work. Brenner disliked the idea. Instead, he tried something called the 'S. B. education society', which involved people in the lab having to explain what their work was about – but at two minutes' notice instead of having a formal seminar. The justification, frightening and invigorating at once, was that anyone who took more than two minutes to get their thoughts together couldn't really understand what they were doing.

And so, says Sulston, however much Brenner disliked the idea of having to behave like a boss, all that side of his job got done very well. 'The boss was there, you know, holding the group very regularly. The seminars happened on and off, but, above all, we were supposed to be interacting with each other and, by golly, we did. It's a matter of getting some reasonably bright post-docs and research students together, making sure they have the equipment that they need, and then being clever and constantly making people feel that there's so much to do, there's so many opportunities and—' Here he broke off as if being once more swept away, and then bobbed up with common sense again. 'In fact, with my personality, the key thing to do was to get away from all that, get away and get something done. But, of course, I don't kid myself, I was being tremendously stimulated all the time by these thoughts coming in.'

Not all the plans came off, of course. Brenner was looking for mutants in the nervous system, and the nervous system runs on neurotransmitters, specialised chemicals which affect the transmission of signals between neurones. They are produced, like everything else in the cell, by small chains of reactions arranged in virtual pipelines which gradually turn a raw material into the finished thing, and it seemed logical that some of the broken nervous system that Brenner was finding in his mutants was caused by a failure in these little chains of reactions. So Sulston was set to try to find how one particular neurotransmitter, known as GABA, is synthesised in the cell, so that they could look for earlier chemicals in that chain. If these precursors were missing in a mutant worm, they would know what it was the defective gene should have been doing. This kind of cross-matching of chemical, genetic and physical evidence was the essence of Brenner's project. That is one reason why the apparently sterile project of mapping the worm's nervous system was so important: with that as a reference, you could see at once if mutant worms were deficient in the way their nervous systems grew into a network. And if the right connections were all in place in a mutant, you would know that there was something chemically wrong again.

Of course, determining exactly what had gone wrong, chemically, in a worm, is not easy. The chemicals of interest are produced for a very short time, in very small quantities. A lot of the time, neurotransmitters aren't to be found because they shouldn't be found. Sulston's first attempts to come up with a test which would show whether mutant worms were deficient in the way they made GABA failed. He never did find a reliable method. Instead, he stumbled into a possibly more interesting technique. A different kind of neurotransmitter, the catacholamines, reacts with formaldehyde to make fluorescent compounds which are clearly visible. No one had managed to make this happen inside so tiny a creature as the worm; but by 1970 Sulston discovered how to freeze-dry worms like Nescafé; except that, instead of little granules of caffeine, he was looking at little granules of fluorescent dopamine. Snap-freezing them ruptured all the cells, spilling their contents into the body of the worm. This was important because it made the cell contents visible. If it were done fast enough, nothing evaporated or diffused, and all the neurotransmitters he was after were left exactly where they had been when they were part of the living economy of the creature, leaving a ghostly glow where its nerves had been.

Treating both mutant and normal worms in this way, and checking all the different stages of the larvae as they grew, gave a very detailed idea of what a mutation in the nervous system might actually be doing. Sometimes, it appeared, the mutation had not interfered with the normal chemical processes of a nerve cell: it had disrupted the growth of the worm so that the cell simply never appeared. But which cell was not appearing? The question led the chemist back to a microscope, and Sulston started examining the living worms to see what was missing. At this point a properly trained biologist might have gone badly wrong, since anyone who had read the textbooks of nematology knew that worms were hatched with all the cells they would ever have. This fact is even mentioned in Brenner's original grant application. But Sulston was a chemist. He didn't know any better, so he looked through the microscope at the ventral nerve cord and looked for nerve cells. What he

found was that there were always fifteen in the first stage
larvae but fifty-seven in the older animals.

Two other workers in the laboratory, Roger Freedman
and Simon Pickvance, had already discovered that the
worm was so completely transparent that individual cell
nuclei could be distinguished inside it when it was placed
under a microscope equipped with special Nomarski lenses
which throw microscopic surfaces into high relief: these
are the same ones as John White melted when he was building his laser scalpel. But Freedman and Pickvance hadn't
done much with this discovery, and they were blamed afterwards by some people for it. Sulston, on the other hand,
spent years, quite literally, bent over the eyepiece; the
sturdy green German microscope still stands on a filing cab-
inet in his office. 'I liked the microscopes and I liked
fiddling with the techniques to get the most beautiful possible pictures out of the microscopes. That was the way
forward and I like to be able to see things. I think visually:
the patterns of the fluorescence enchant me and the patterns through the Nomarski microscope – they're absolutely
entrancing.'

He tells the story without any eureka moments. 'The
artisan trick there was to realise that I shouldn't try and kill
them, which is what everybody else had tried to do: they
tried to paralyse or immobilise them, and I, for some
reason, gradually came to the realisation that you could
actually watch the worms as they moved, provided they
were well fed. So it was a matter of getting the worms
nicely mounted on a little pad of agar under the microscope with just a scraping of bacteria in the middle which
they could eat, and they would just stay in that little zone of
bacteria. They wouldn't stray out and get lost at the edge of
the slide and, although I had to be constantly checking the
microscope, I could watch them.'

Three of the most important facts about the worm came together under his eyepiece: greed, transparency and fast growth. He placed a newly hatched larva on a dab of bacteria between two slides, focused on the ventral nerve cord, and settled down to see if any of the cells he was watching would divide. What actually happened seemed to be a great deal odder: ten fresh cells appeared among the fifteen he was watching, apparently by spontaneous generation. Watching the cell division in a living animal would have been a landmark in biology: but spontaneous generation would have been epoch-making. He looked again.

It took a little thought to work out what had happened: the Nomarski lens gets its effects from having a very shallow plane of focus. This makes anything within the plane stand out like a bas-relief; but he had been focused on a few cells deep within the transparent animal. If the new cells had originated a few microns deeper into the animal – or indeed closer to the surface than the ones he was watching – they would have been invisible until they pushed their way in from elsewhere. The animal was growing so quickly that the cells in it moved around while he watched. He kept watching. The cells that had so mysteriously appeared started to swell and divide.

This was wonderful in itself: he found himself watching a living animal at exactly the place where it was growing. Cell division is the central process of development: without it we'd all still be fertilised eggs.* But to watch it in a living animal is something new; and he realised that he could in principle watch the whole growth of the worm at a cellular level, and identify exactly the lineage by which each cell in the finished worm descended from the original egg. All that day he watched the ventral nerve cord grow. At the end of

* Assuming chickens.

the day, reluctantly, he left the lab, but he took his worms and microscope home. He had originally planned to keep on watching the worm, checking every ten minutes to see if anything had happened, for the whole of the thirty-six hours it would take to grow up completely. But he realised some time in the watches of the night that this would be impossible. So he gave up and put the worms in the fridge* – 'They weren't happy about it, but they survived' – and in the morning he was watching again. As he watched over the course of an extraordinary weekend, each of the ten cells he'd started with divided until it had had six descendants.

What was even more wonderful was that the new cells all seemed to behave in the same ways, as if their fates were determined by their genes. Among each group of six descendants, one turned out not to be a nerve cell, and migrated to another part of the body. The remaining five all became nerve cells on their own, making sixty he was watching. Four moved away from the plane of focus: this movement of cells slithering around each other is the factor that shapes our bodies as much as cell division. Another four of the new ones simply died. This in itself was significant, because it was always, in every worm, the same four cells that died. They had been instructed by their genes to do so just as much as others had been instructed to become nerve cells or muscles. It was the first time anyone had seen this programmed cell death in the worm; along with division and migration it's the last of the three processes that make all living bodies, including our own.

The story as I've told it may seem a little bathetic. It could be told with trombones and kettledrums instead. You

---

* Under the lettuce: it was too cold outside the vegetable section.

might say that Sulston, peering down the microscope, had just won his Nobel prize. You might say, too, that he had seen everything cancer researchers would be watching for the next thirty years. Cancers, in humans as in other animals, are what happens when the processes of cell division and cell death go wrong. The cancer cells divide too often, and they don't get or don't respond to the signals from their neighbours telling them to die. But telling the story that way would be untrue to everything but the facts. Sulston was not trying to make history, or even to benefit humanity. He was driven by forces deeper and more powerful. He wanted to tinker, and to understand.

Finding things out in the worm was an awful lot like being in love. The state of delighted, watchful patience that Sulston reached, lost in the new world at the other end of the microscope, seems to me the same as the way that moonstruck lovers gaze at each other. They just want to look, to drink in the presence of the beloved, to learn everything there is to know.

Sulston took notes as well. There are pages and pages of notebooks now, filled with neatly coloured circles and lines, showing the differing lines of descent of every cell within the worm. But that first weekend, he mapped only the cells of the ventral cord, as the larvae grew to adults. When Monday came, he was able to go to John White and try to find out what the cells were that he had discovered. White had by this time mapped the ventral cord completely, using the slices Nichol Thomson had made: the really time-consuming and difficult piece of his task was to be the mapping of the nerve ring round the pharynx. On the ventral and dorsal cords, so little happened at a microscopic scale that Eileen Southgate could often afford to skip as much as .02 of a micron as she worked her way through the pictures of the worm.

On the basis of these maps White had already distinguished seven different types of neuron within the bundle of fifty-two that made up the ventral nerve cord. Now he settled down with Sulston's new data to find out where in the worm these different types of cell had sprung from. What he learned was that the fate of a cell was determined largely by who its ancestors had been. This was an answer to one of the questions that Brenner had set his team to think about: whether the society of worm cells was European or American. In a European society, your class was determined vertically, by your ancestry: if your father is a peasant or a gentleman, you will be one too. In an American society, Brenner explained, social relationships are determined horizontally, you are a member of the upper classes if you are rich enough to have other rich friends, and never mind who your parents may have been.

The neurons that White and Sulston had teased out stuck to precedent so closely that they might have been listed in the *Almanach de Gotha*. In fact they came to be known by their ancestors, like German nobility, so that a cell may be identified as P.apa, meaning that it is the anterior daughter of the posterior daughter of the anterior daughter of the Precursor cell.* White soon discovered that among the descendants of the ten new cells that had originally swum into Sulston's ken, all the P.apa cells grew to become a specific type known as a dorsal AS neuron.

Arguments over whether all cells should be European and American left a lasting mark on the building: at one stage White and Brenner had a bet about the way in which

---

* Animals have two sorts of 'back': the end that is opposite the front, and the side that is opposite the belly. So to avoid ambiguity the front/back pair is known as anterior and posterior, while the belly/back pair are ventral and dorsal.

one particular clump would develop; the stake was a bottle of wine. When White won, Brenner produced it in the seminar room, but no one had a corkscrew. What they did have was a great deal of scientific equipment, including an aerosol of freon gas for blowing the dust off microscope lenses. It was the work of a moment to hook that up to a hypodermic needle, thrust the needle through the cork, and then extract it with raised pressure. At one stage corkscrews were sold that worked on these principles but they were pumped by hand. The White–Brenner gadget was far too highly pressurised, and when the cork flew out of the bottle it was followed by a geyser of red wine. It left a splash mark on the ceiling which was there for years.

Sulston by now had a slightly ambiguous position in the labs. He was a full-time member of the LMB staff, but on a level slightly below that of the other ones, though slightly above the post-docs. He did not think he was going to bring huge glory to the labs by publishing, and neither did Brenner. But he loved his work and was good at it, so he was found a post where he didn't have to do much of anything else. It seems entirely in character that this arrangement delighted him because it meant he could talk to everyone and be responsible for hardly anyone. Judith Kimble, who came later to study as Sulston's post-doc, remembers him sitting in his room all day over the microscope, leaving her to get on with her own work. He loved his perch on the mezzanine floor of the organisation. 'That was wonderful because I was free to go on interacting with everybody and, of course, I did.'

Still, when Brenner went back to South Africa in 1972 for the first time since emigrating in 1956, there was real concern that as an outspoken left-wing Jew he might get into trouble with the apartheid regime. The man he nominated as his successor on the worm project was John White, not Sulston: 'If anything happens, John White does my worm stuff,' one observer remembers him saying.

Once Sulston had shown that it was possible to determine cell lineages in the living worm, there was almost a race to get as far as possible with the process. Formally, Brenner's 1974 paper on the mutants of the worm marked the point when the project could claim to have established its credentials. But the excitement had started to grow before then; Americans who had come over to work in other parts of the MRC began to work on the worm instead. Bob Edgar and David Hirsh, who had come to work on tRNA, which helps to assemble proteins, both moved up a floor at the MRC to help with *C. elegans* and then returned to America having seen the light that shone through the worm. But the technical breakthrough that Sulston had made when he first watched cells divide with the living creature was first savoured by two younger Americans, Bob Horvitz and Judith Kimble.

Kimble had by that stage returned from Copenhagen to Boulder, Colorado, where David Hirsh had a letter from Sulston telling him of the new technique, and suggesting he might want to map the gonad. He thought this was a suitable project for his graduate student, and so Kimble got her doctorate by watching, as Sulston had done, the cells of a worm dividing beneath the Nomarski lens; but she was watching the development of the gonads – she did both male and hermaphrodite – not the nerves. At the end of the process, there was a bent tube within which the eggs would develop – you could watch this – and move slowly along and round the bend until they were fertilised, and then filled the belly of the worm. Within the fertilised eggs, glowing like faint blue bubbles, could be discerned the curled-up embryos of worms as they developed towards the larvae that would hatch. But no one, at that stage, thought it would be possible to watch the embryos growing.

Horvitz arrived on Guy Fawkes' Day 1974, just before the great wine-bottle explosion. The most recent part of his career had followed almost exactly Brenner's trajectory from phage genetics to neurobiology: he had been a graduate student under Jim Watson and Wally Gilbert (a former physicist who later got a Nobel prize for discovering a way to sequence DNA). 'I grew up in the cold room,' he says, 'doing biochemistry and applying genetics to the problems of the development of the phage.' The period before he 'grew up' had been eventful. He was raised in the Midwest, and was about to go to the University of Illinois when John Kennedy was assassinated. Somehow this made him think he needed wider horizons, so he went instead to MIT, where he picked up two degrees, one in theoretical maths and the other in economics. He also worked as a computer programmer in the summers, for IBM. 'I was involved in designing the first conversational computer programming system, which they never released.' 'Conversational', in this context, meant that you could use the computer without ever learning to program. So he was engaged on the work that means we now think it quaint that Brenner and White had to wrestle as they did with the Modular 1. However, IBM never released the programs; and in any case Horvitz thought the most interesting thing about his undergraduate years was neither his degrees nor the computing work but the journalism he did in his spare time: he was managing editor of the MIT newspaper, and 'I spent most of my time at the *Cambridge Chronicle* being impressed by the vast quantities of Scotch these guys drank as they were putting together the newspaper. But I spent two nights a week overnight putting the paper out. And I was interested in a variety of other things.'

Now he is a neatly trimmed man, drawling seriously in a spacious and comfortable office at MIT. But in the Seventies he had the hair, the sideburns, and the heavy-framed glasses of the serious radical. The whole lab was full of counter-culture in the early years, like everywhere else.

One of the reasons he went into biology was that he felt that economics was too remote from the real world. He wanted to change big things. He wanted to understand the brain. He thought that with a degree in maths and one in computation he was most of the way there, but a rigorous training in biological research was exactly what he wanted. Rigour – and excitement – was what he got. The genetic code had been fully worked out only in 1966, and he started in graduate school in 1968. Molecular biology was growing in ways that made it hugely exciting, and almost incomprehensible to an outsider, even one as clever and energetic as Horvitz. 'I can remember sitting in courses in my first semester and I had no idea what people were talking about. I can remember thinking, "If I don't have a better idea at the end of the year, I'm doing something else," because I didn't think I had a chance in this business.'

He struggled on. By the end of the year, not only had he begun to understand what his lecturers were saying, but he found it was fascinating. The phage he was working on were further from having a nervous system, and had less behaviour, than even the bacteria they preyed on. If he wanted to approach the problem of a brain he would have to work on an animal, and the choice was between the fruit fly and the worm. The fly was more respectable, in scientific terms. But the worm seemed to him far more interesting, for all the reasons that had appealed to Sydney Brenner. Horvitz, like Brenner, wanted to hack at a nervous system with genetics, and the tiny, prolific worm had to be better for that than the large and relatively slow-breeding fly. That seemed obvious to him, even though there had not been a single paper published on it then, despite six years of work. Applications for a post-doctoral post are meant to be garnished with references to the scientific literature which show the student is entering a fruitful and interesting research project. Horvitz had one reference on his, and that was a 'personal communication', scientese for gossip.

He had never met Sydney Brenner, nor been to the English Cambridge, but the woman he was living with had toured England earlier, and assured him that it was a beautiful city. That did it. He wrote to Sydney, found a post, and the couple crossed the Atlantic to his new job. As they got off the train, his girlfriend looked around and announced that she had never seen the place before. She'd been thinking of Oxford all the time. Despite this setback, Horvitz thrived.

'There was a whole new world open because no one had ever touched the organism before. So you could do anything and find something new. That's very gratifying for a scientist because so much of what goes on in science, you're doing it and afraid that somebody two doors down or across the country is doing exactly the same thing so you gotta do it faster. Well, in this case there was nobody to bump into. Nobody was doing this, so whatever you did was a discovery.'

Almost at once, he found himself collaborating with Sulston on the lineage work. The two men make an extraordinary contrast in temperaments, who seem to fit together very well. It is not that they are so very different in terms of drive, but their ambitions appear poles apart. Horvitz is disarmingly modest and mild and it takes an effort, listening to him tell the story of his life as if it had been an engaging shambles, to realise that here is a man who got two degrees at MIT in four years while editing the student newspaper, and then waltzed into a job in a lab run by two Nobel prize-winners before going on to get one himself. But Sulston's humility is something else again. As it happened, both of them have ended in the front rank of their profession, sharing their Nobel prize; but I think Sulston would have been perfectly happy to end up just behind the front rank, whereas Horvitz would not. Sulston looks down at particulars; Horvitz looks up at horizons. When I first talked to Sulston about this book, he said happily 'You've got to talk to Bob Horvitz. He's spent thirty years studying twenty-two cells in the worm's vulva,' and he meant this as a tribute to the inexhaustible riches of the worm. Horvitz, by contrast, talked about how he had studied behaviour and development on parallel tracks ever since he started with the worm. Both men were telling the truth. It is just a difference in style which made them work well together.

When, later, I teased Sulston about his joke, he said, 'He studied the vulva because all life is in there. The problem of life, at the moment, is still to understand the cell and the way cells interact. It's a huge problem still, and that's exactly what he was studying in the vulva. It doesn't matter which cells you choose, so long as they're well organised and you get many footholds into them and the worm vulva was a nice . . . . And, of course, again, there was not a vulval plan. What there was were a lot of mutants that turned out to have significant reproducible effects on the lineage of the vulva. That's what got him in.'

It may have helped that they were both entering a new field, like almost everyone else in the lab: Horvitz says, 'I was a mathematician, who wasn't really a mathematician because I was never at that level, and he was an organic chemist and there we were, trying to solve the development of a worm. John White was really a physicist and computer scientist. I think the only one who was respectable was Jonathan Hodgkin: because he was younger he actually got a degree in biology.' In any case, Hodgkin got his doctorate and went off to California to look at bacteria three days after Horvitz arrived and inherited his space in the lab.

He was not at first enthralled by what was being done there, and the reason provides a delicious insight into the confidence of the first molecular biologists that they could explain everything in nature without too much reference to traditional knowledge. At Harvard, in Jim Watson's lab, Horvitz had done 'hard science, with radioactivity, gels, and molecules, and [I] had to be convinced that what my eyes could see directly provided scientific information as reliable as what [they] could see when they viewed the output of a scintillation counter.'* But once he started look-ing through the microscopes, and discovered that what he saw was not only thrilling, but elegant and suggestive when mapped, he was away.

Sulston didn't at first believe they could map the whole worm, but Horvitz egged him on to do more of the lineage. The two men started a kind of relay race through the worm which lasted for about two years. They worked hugely long hours. The logbooks of their work are simple exercise books in which the different cells are marked by colour-coded circles. Horvitz pulled down a stack from a cupboard in his office at MIT and showed me an entry. 'As each cell started to divide, the time would be marked beside it: here is one that starts at 8.15pm, and by 8.25pm, here are two cells.' The last entry for that night was made at 12.41am, and that was by no means exceptional. When the two watchers were exhausted, they would put the worms in the refrigerator for the night.

* R. Horvitz and J. Sulston, 'Joy of the Worm', *Genetics*, vol 126, October 1990, pp. 287–92.

They also wrote to David Hirsh, in Boulder, suggesting that he do some work on the gonad. He passed the job on to a graduate student, Judith Kimble, because she was more interested in cell biology and development than in the other areas where his lab worked, biochemistry and the early attempts at gene cloning. 'She was a good student, she was curious, and she was in love with the idea of doing development. We all had that interest, but you had to find the student who had enough to develop that patient way of charting lineages.'

She wasn't sure herself that she had: it seemed a huge amount of work for a graduate student. But she spent her eighteen months hunched over a microscope, and that was what gained her entrance to the LMB. The hermaphrodite gonad develops almost entirely in the fourth (L4) larval stage, since mating is the only thing a grown worm does which a larva need not do. It starts as a little blob of cells in the middle of the worm, and then the cells at each end start to move out towards the front and back of the creature, trailing a growing gonad tube behind them. Finally, these distal tip cells change direction, and grow back towards their starting point after making a hairpin bend; the cells that follow behind them divide to form a tube. When the distal tip cells have almost reached each other again, they form the spots where the gonad starts to manufacture egg cells, which move slowly back down the tube until they reach the sperm which has developed at the far end, where the gonadal blob originated. It's an extraordinary and intricate dance, all over in about eighteen hours.

Most of the lineage work had to be done, in one sense, backwards. What they would really have liked to be able to do was to look at a cell and tell where it had come from, but that was impossible. You could not play the tape backwards. Instead, they had to make guesses about the ancestry of cells that interested them, pick a candidate ancestor, and watch what happened to it. Where, for example, did the cells of the vulva originate? The point about the vulva, in this context, is that newly hatched worms do not have one. It is one of the things that they must grow as they moult into adults. This is development, but at the same time the contractions of the vulval muscles, when they expel the egg, count as behaviour.

Most of the muscles the newly hatched worm puts on as it grows are easy to trace. One cell, rapidly named M, located towards the front of the worm, divides into eighteen muscle cells in the first few hours of the worm's development. According to classical embryology, these must be the cells which, on European principles, give rise to the other muscles cells of the vulval wall, when they appear, a long way away in the worm, and several moults later. That was Horvitz's view: Sulston upheld an American pattern of development. Surely, he argued, the vulval muscles would descend from cells in the gonad, which were very close physically and not very different.

So each man started a long stint at the microscope, watching and watching different worms as they grew, tracing the fates of their candidate cells. Horvitz turned out to be right. Though Sulston's cells proliferated into all sorts of things that might become a vulva, it was sixteen of Horvitz's muscle cells which finally set off on a journey around the worm to become the muscles that lined the vulva and expelled the eggs. These muscles were important in two perspectives. They were, obviously, an example of development. But they were also an example of behaviour, since egg-laying is undoubtedly a behaviour. Armed with their cell lineages, Sulston and Horvitz could start examining mutants of movement to see whether any of them were really the result of failures in cell division.

There seemed to be nothing they could not explore. But, apart from a conviction that the genes must in some way determine what they were seeing, they had no detailed theories.

As they thought about the problem, the two men grew less optimistic. Though the regular pattern of cell division made it clear that this must be under genetic control, it seemed perfectly possible that any disruption among these genes would have effects all over the animal's body. They reckoned there were three areas in which they might find mutations affecting cell lineage. The first was egg-laying. It is a behaviour, but one which develops through successive larval moults. The newly hatched larva has all the nerves and muscles that it needs to be able to feed, but none that it needs for egg-laying. If they could find worms whose egg-laying was disrupted, they might have found the effects of a gene controlling the development of behaviour.

As it happened, he had one mutant which was perfect for the purpose: the disgustingly named and depressing to contemplate 'bag of worms'. It is not too uncommon: by looking at about 15,000 mutagenised worms he found fifty-seven bags of worms, which are very easy to recognise, and look, under a microscope, rather beautiful, like a skein of embroidery silk. Later he found that he could get them more simply by centrifuging a mass of worms in a liquid of carefully judged density: the normal ones formed a clump at the bottom and the bags of worms floated. They get their name because in such worms the vulva fails to develop at all. Since the worm is a self-fertilising hermaphrodite, this doesn't stop it laying fertilised eggs, but the eggs can't escape. Instead, they hatch within the parent's body, and with nematode pragmatism eat it alive, from the inside, until they can burst out. But because this is a genuine, heritable mutation these worms in turn have no vulvas, and will be eaten by their own children. The bag-of-worms line supplied Horvitz with an endless supply of animals with which to pursue his investigations into the growth and development of muscle.

The second category of mutants they wanted to explore were those which moved oddly because the ventral nerve cord – the first part of the animal whose growth Sulston had mapped – had not developed properly. It seemed perfectly reasonable to suppose that an animal whose nerves had not grown properly would not have its muscles under proper control. If this failure to move persisted down generations – in other words, if it was a genuine mutation – it might be the result of a gene breaking which affected the division of the nerve cells in the ventral cord.

Finally, in case any of these theories did not pay off, they also screened animals at random. They would look at one worm in a clone, and keep the rest going as spares for further investigation if they found anything interesting, because their tests were made on very dead worms.

They certainly couldn't just watch animals grow under the microscope in the hope that they would happen to be watching the interesting part of an interesting mutant just when it did something unexpected. The two men had taken years, and thousands of worms, just to establish the normal patterns of cell division within a cell. To try to watch this in real time, at a microscopic scale, might have taken decades.

Instead, they started to analyse mutants, to see if any of them were abnormal because they had the wrong number of nerve cells. If it turned out that they did, that defect might very well be the result of a failure of one of their common ancestors to divide at the right time; and now that Sulston and Horvitz had the family tree of every cell in the adult worm, they could trace back through the interesting mutant to the right stage to see if that was its flaw. If they found one, they would know they had discovered a gene for development.

The first thing to do was to go over once more the mutants using the techniques that Sulston had first developed when he came to the lab, which made the nerve cells fluoresce if they contained particular neurotransmitters. No fluorescence probably meant that the nerve cell they wanted was absent altogether, but it could be spotted quickly under an ordinary microscope, since you did not have to focus on the precise plane of the nerve cell. Judith Kimble, who was doing the gonad's cell lineage at about this time, remembers that the operation of the Nomarski-lens microscope required constant two-handed adjustments as you followed particular cells as they moved around within the worm, which was itself moving as it munched bacteria. Fluorescence, shining through a transparent worm, was a much quicker signal that there was something worthy of closer investigation.

By using all three of these techniques they managed quite rapidly to find twenty-four different cell lineage mutations, or twenty-four different ways in which a worm might grow to maturity with the wrong number of nerve cells, and pass this fate onto its progeny. Crossing the different mutations with each other – using males, of course – established that there were fourteen distinguishable genes involved, some of which might have several different effects.

Until then, the triumphs of other worm projects had been technical and practical. People had learned an incredible amount about the worm. This was the first time the worm showed something important and unknown about genes themselves, and the way they worked. It was the moment when the worm started to make the contribution that Brenner had planned for it, not to knowledge about the worm but to knowledge about the world. Through the transparent body of the worm, they had found not only cells in an individual animals but the workings of the genes in every animal. It had become, as it was meant to be, a lens through which all the animal world could be seen.

6

Embryonic Lineage

Underlying this progress was the discovery that worms and humans are assembled from the same proteins. Some of the genes responsible for cell death are the same in worms and humans: it was for this discovery that Horvitz won his share of the Nobel prize; but it took him some years to find them. One of the first significant *unc* mutants was *unc-54*, which made worms that couldn't wriggle because their muscle was malformed. A worm whose copy of *unc-54* is defective cannot make one of the strands of proteinaceous macaroni which go to make up myosin, an essential muscle protein. Muscle, in all animals, works because strings of proteins slide past each other, using myosin and actin to do so. Myosin is a hinged protein, with its base embedded in one fibre and a hinged arm sticking out to engage in the actin on the next strand; when the hinge folds up, the two strands of muscle fibre are pulled an infinitesimal distance past each other, then it opens again to engage in the next molecule of actin. Repeat this process an almost infinite number of times, and you have the muscles of Arnold Schwarzenegger bulging as they contract. Repeat it many fewer times, and you will get the sinuous writhings of a worm. In both cases, the underlying mechanism of actin and myosin is the same, even though the muscle fibres in vertebrates are arranged and attached quite differently from those of *C. elegans*.

Worm myosin was discovered by Jon Karn and Sandy MacLeod, and then worked on by two American post-docs at the LMB, Henry Epstein and Bob Waterston. Waterston had been an undergraduate engineering student before he switched to medicine and qualified as a doctor. None the less, he seems to me an extreme example of the biologist who sees the worms as interesting Meccano, rather than as animals. He was the first American post-doc who came specifically to work on the worm: he had been converted by a lecture Brenner gave at Woods Hole in 1969. 'I was interested in academic medicine and I'd learned more about it and got into immunology, but I found as I was going through courses, particularly during my PhD period, that genetic approaches were just that much more appealing to me. They were more defined – and when Sydney talked about the worm it just seemed like a system that was complex enough so that all the central questions weren't already answered, and yet defined enough so that you could really hope to get your mind around what happened in development.'

John Sulston had discovered the chemicals that would break open large numbers of worms, leaving what was mostly muscle behind: individual worms were far too small to extract the muscle from, but done in bulk, and using the techniques developed by other LMB researchers, Epstein and Waterston were able to check that worm myosin reacted properly with actin from other animals and vice versa. So they really were different forms of the proteins, with a common ancestry. The *unc*-54 mutation actually worked to shorten a part of myosin in ways they could understand.

'That was the fun,' says Waterston, 'the thing that really drove it, and made it all tangible: we could detect the shortened myosin even with the crude tools that we had in those days. So we were able to start correlating genes and proteins in direct fashion.'

This was something people had been hoping to do ever since the 1950s, when it was established beyond doubt that genes made proteins. It seemed to Waterston that the worm became the lens in a pin-hole camera, through which the whole of biology might be visible. Now they had found a gene and were able to see exactly which protein it made, even though it was all done without any sequencing. It was to be nearly ten years – 1983, in fact – before he did his first sequence of a gene, using methods just worked out by Wally Gilbert. It took him perhaps a month to get a gel recording several hundred bases of the gene he was interested in: it looks like a great square barcode, showing the different lengths of gene ended by a particular letter. Three of them wrote down the key – which letter was on which stripe of the barcode – and they puzzled out by hand what is now done in fractions of a second by a computer. This painful process got them where they wanted to be: they were able to see that in their sequence there were the three letters (AUG) that told the cell to stop making a string of amino acids. In this case, just as with *unc*-54, this triplet, the stop codon, came too early, making the thing what is called a nonsense mutant. That was a wonderful moment. 'It was really spectacular and primitive – just so primitive!'

But this is jumping ahead of the story, towards the point where massive sequencing techniques have made it so obvious that the worm and the human are made of very much the same things that a researcher can mention, quite by chance, in a lecture that it is really easy to forget whether he is inside a mouse or a worm or a human sequence, and no one will laugh except for one slightly shocked journalist at the back of the room. Back in the late Seventies, there was one more piece of work to be done before the canonical picture of the worm could be established; and it was this that made John Sulston's name and established him as the man who would clearly inherit the moral leadership of the wormy bits of the LMB from Brenner.

By 1979, the whole of the worm's lineage had been traced from the moment of hatching. The newly hatched larva had 550 cells,* and each and every one of them had been followed through Nomarski-lens microscopes by Sulston, Horvitz, and Judith Kimble and its fate established. But how did those 550 cells arise in the larva? It was such an obvious question that people had been trying to answer it for years. Because the worms are transparent, and the eggs inside them transparent, too, the embryos could be seen developing inside their parent even in a normal egg. But everything happened so fast, so formlessly, and in such a small space, that it seemed only a video camera could possibly capture it. A German group at the University of Göttingen under Günter von Ehrenstein had been trying for years to automate the process. In the Seventies, they kept issuing announcements of imminent progress: in 1975 after two years' work they had managed to establish the worm's history through the first three cell divisions, and announced† they had bought a really tremendous computer to help them with their work.‡

The next year, they followed the embryo through another few rounds of cell divisions, up to 170 cells, and their computer was upgraded still further. This looks very

---

* *C. elegans II* (New York: Cold Harbor Spring Labs Press, 1997).
† *Worm Breeders' Gazette*, 1: 1.
‡ A PDP-11.

much like White and Brenner's original plan to automate the slicing and display of the worm: the Göttingen group were slicing eggs for electron microscopy just as Nichol Thomson had sliced whole worms, and 'One ultimate goal is to display the important events in embryogenesis in a highly schematised movie on the screen of the graphics display,' said their announcement in the *Worm Breeders' Gazette*, a magazine which had been founded to hold the burgeoning international worm community together.

There matters seemed to hang: the speed and complexity of what went on in the egg defeated even the highest-speed video cameras. The Göttingen group had reached a stage where there were 170 cells formed into a tube quite like a worm which were dividing simultaneously in half an hour, and each one needed to be traced. The technology just couldn't keep up. By 1977 they had got up to 182 cells and had found a way to remove eggs from the parent worm and get them to develop under microscope slides; but there matters seemed to stall.

Horvitz, as usual, encouraged Sulston to have a go himself, using the manual techniques that had worked with the rest of the worm. Sulston was reluctant, as usual, to start work that looked as if it would tread on someone else's toes. 'They were doing it before I was and I hate to duplicate work: I hate to compete needlessly. There's always plenty to do and it seems stupid to wade in on somebody else's patch.'

This sounds like common sense, but it is also an attitude which appears to be pretty uncommon in modern science, where competition for resources is alarmingly ruthless. Remember that one of the things that had attracted Horvitz to the worm in the beginning was precisely the absence of competition, which suggests that it was an inescapable feature of research elsewhere. Even Crick and Watson's unravelling of the structure of DNA was seen by Watson as a race – and that's how he told the story, too, even though Crick never saw it that way. Judith Kimble said Sulston delayed doing it simply because he was 'so incredibly nice'; Sulston himself suggests earthier motives: 'Of course, I also didn't know if I'd be successful so I can't claim any great altruism. It was partly a matter of personal feelings and not wanting to be unpleasant and partly just practicality. [The embryonic lineage] was going to be a big investment – years and years. But, obviously, they were just not getting anywhere so I just started in, in the end, and did it once I found I could do it.'

In fact, it took him only eighteen months to establish what was happening in the fourteen hours it takes a worm's egg to hatch. The crucial technical trick was not just going back from computers and automatic recording machines to one man at a microscope: it was going back to the tricks of traditional astronomers, who used cross-hairs made of spiders' webs to help them focus on particular faint stars. He needed to focus on one cell in the embryo at a time. Almost all the cells looked very similar, and they were all crammed into a tiny space, and moved around within the growing worm even more than they did after it hatched. So he needed something to help him find his way through the embryo by dead reckoning. You can buy cross-hairs from Zeiss – who made his microscope – but they turned out to be worse than useless for the incredibly detailed work he was doing: because the sighting cross-hair was engraved into the glass under which his specimens rested, any muck on it got magnified and distorted. So he made his own gossamer crosshairs, and with them was able to have a firm sight on the cell he was interested in, even if he looked up from the microscope for two or three minutes.

After that, it was simply hard grind. The Germans, after four years, had mapped the embryo up to its twenty-eighth cell division using a whole team of students and all the technological resources they could muster. Sulston did the rest by eye, alone in his tiny cubicle. He settled rapidly into a routine of working two four-hour shifts a day with a concentration that awed the rest of the chatty lab, though they worked notionally much longer hours and certainly more varied ones. For the post-embryonic lineages he had done earlier with Bob Horvitz, a different sort of application had been needed. Because nothing might happen there for half an hour at a stretch, what was needed was a willingness to work long hours, and, after the first thrill had worn off, a certain tolerance for boredom. The embryonic lineage placed on him 'exactly the opposite demand: things were going too fast; all the cells looked alike. On the other hand, it was all over quite quickly so one had to readjust, and the way I readjusted was to become very, very disciplined about it. I absolutely excluded people from the room and refused to talk to anybody. I just couldn't be interrupted at all.' Judith Kimble remembers an accident in the lab filling the place with vile-smelling fumes. Everyone left for coffee until the smell had gone away, but when they came back, there was Sulston, still in his tiny room, still bent over the microscope as he had been the whole time, ignoring everything that was not beneath his lens.

He even abandoned the appointed rite of coffee during this period. Phil Anderson, then a post-doc learning worm genetics after training on bacteria, remembers that, 'Whereas post-docs were often yakking it up in the coffee-room like this, talking about various things, John would be in his microscope room, watching larvae develop and trying to describe the complete cell lineage of the animal. He would interact but always kind of in a cerebral, detached manner, so I found him a little distant. He'd come out every now and then to get a cup of coffee but, whereas we'd all sit down and yak it up for an hour or so, he didn't have an hour because the cells were dividing, he'd got to get back and see what happened.'

Anderson's recollections make an interesting contrast with Sulston's view of himself as someone who was able to wander around the labs talking to anyone, and who was not, by the exceptionally high standards of the MRC, himself exceptional. 'When Sulston would present information, you just knew you didn't need to see the data behind a statement. He would say that "This cell mediates this process". With a lot of scientists, when they gave a conclusion, as opposed to an observation, most other scientists would ask "What's that based on? How good is that conclusion?" but with John, you just knew you didn't have to do that. With John, you just knew that it was locked solid tight. Very impressive in that way.'

Everything was double-checked with Nichol Thomson. Once Sulston thought he had traced a cell, the eggs were sliced just as the whole worms had earlier been, and the embryos knotted like pretzels within them examined beneath an electron microscope. All his earlier lineage papers – since the very first one proving that the ventral cord of the worm could be watched developing and the individual cells traced while it happened – had been published with collaborators. His first lineage work he published on his own, though it appeared with one of John White's, because he reckoned his discoveries were useful only in collaboration with the map of the nervous system; later work was published with Horvitz, and the biggest paper, on the embryonic development, was published with Einhardt Schierenberg, who had done much of the work at Göttingen (poor von Ehrenstein had died of a heart attack while out jogging on Boxing Day 1980), John White and Nichol Thomson. But it was Sulston who won instant recognition for that paper. Brenner put him up as a Fellow of the Royal Society and he was elected at once.

'I didn't actually realise how much people would appreciate it at all. I got one or two awards, along with Bob Horvitz and people, for it all but I didn't think that people would think there was any great theoretical significance in it. I didn't really . . . you know, I thought it was fun but it was just a map, really. But it transpired during the Eighties that people thought this was important.'

For a while after that, he didn't know what to do. The logical step, now that he knew far more than anyone else in the world about how the early development of embryos should proceed, was to go back to genetics, and to try to find the mutants where development was disrupted within the egg. That, or something similar, was what the remarkable German Christiane Nüsslein, also at Göttingen, had done with fruit-fly embryos, though she later changed organisms to zebra fish. Mutants which killed an embryo even before it could hatch must by definition reveal extremely important genes which could then be examined with all the other tools that were becoming available.

But Sulston discovered, rather to his surprise, that he no longer enjoyed genetics very much. There is monotony and monotony in research; and counting worms through the generations seemed to him an unrewarding form of jigsaw puzzling. In the early Seventies, when he had mapped the formaldehyde fluorescent mutants and learned genetics in the process, he said it was huge fun.

It's clearly a form of fun that appeals to some temperaments more than others. Mapping genes involves discovering how many generations it takes for two mutations to diverge, which is a matter of counting the number of worms with each mutation in every successive generation for a month or two. A good worm counter, says Sulston, can do a hundred a minute, or 3,600 an hour; and in the Seventies this was done by collecting the worms of interest with blobs of sticky bacteria on the end of wooden tooth-picks, for hour after hour. Later, Sam Ward, one of the earliest American worm researchers, invented one of those gadgets which make the world look entirely different: he fastened a dissecting needle into a miniature soldering iron and then ran an electrical circuit through the soldering iron and the petri dish in which the worms lived, so that it registered and counted each pulse of current as a worm was fried. The announcement of this breakthrough in the first issue of the *Worm Breeders' Gazette* is an interesting scientific communication. 'In genetic crosses and in experiments measuring the number of progeny produced by a worm, the counting of progeny is tedious. It is often important to kill off or remove all of one generation before they reach maturity in order to prevent a second generation from confusing results. We have found that a miniature soldering iron can be used for killing individual animals on a petri plate. The killing is reliable and fast if the tip is cleaned of roasted worm periodically . . . The whole system can be purchased and assembled for about $60.'*

The ease with which large quantities of worms could be obtained was one of the things that had made them attractive in the first place. None the less, when confronted with a recipe in the *Worm Breeders' Gazette* for breeding males which starts, in effect, 'Take 40,000 eggs', it is possible for the stoutest heart to quail. It is also possible that Sulston felt he would like to straighten up from the microscope for a few years. Muriel Wigby, recounting her decision to retire,

---

* *Worm Breeders' Gazette*, 1:1.

says, 'After I had done about thirty-three years in a lab, I began to think, "If you saw a little dog tied to a stake all day, you would think that was cruel. So what am I doing stuck on a microscope all day?" I slipped my lead and went home.'

Sulston did not want to retire. But he refused all suggestions about mutants, and started to think, instead, about a map of all the genes then known on the worm. Traditional gene maps said very little about the underlying DNA. They merely showed which genes lie next to each other, and on which chromosomes, knowledge acquired by slow and painful observations, of the sort that had made Muriel Wigby feel like a puppy tied to a stake. What Sulston had in mind was much more ambitious. He wanted to combine that traditional knowledge with the new understanding of DNA that was emerging as people learned how to break up a genome by its chemistry rather than its function.

It was not just ambition that drove him. He also had a strong feeling, by the beginning of the Eighties, that the methods that had carried them so far were getting exhausted: 'What was very apparent in the worm was that people were absolutely log-jammed about getting the genes isolated.'

It was nearly ten years since Bob Waterston had found the gene that made the *unc-54* mutant, and since Brenner's paper laying out everything known about the genes that could be detected by the visible effects of their mutation on the worm's development. Most of what could be found out by those means had been discovered, Sulston felt; and it wasn't ever going to be enough. There were just too many effects a single mutation could have on the visible development of a worm, and the horizon of hard-edged, digital knowledge that made molecular biology attractive seemed once more to be receding. They were going to have to get down from genes to their constituent DNA. 'We knew the pleiotropy,* we knew it was all terribly complex. People had begun to think, "We've got to have the sequence, we've got to find out what the molecules are, what they do." We were getting more reductionist, in a sense.'

* Pleiotropic genes have an effect simultaneously on more than one part of the body that contains them.

7

The Worm Goes West

Bob Edgar met Sydney Brenner in the summer of 1954, when the two men shared a lab bench at a summer school in Cold Spring Harbor on the care and manipulation of phage. The phage – viruses which prey on bacteria – are so simple that it is reasonable to ask whether they are living at all. They are merely strings of DNA wrapped in a protein housing which looks like a moon lander, and they can only reproduce when they have broken into a bacterium which contains all the chemical machinery needed to copy DNA and make proteins from it.

This simplicity made the phage ideal for the first generation of molecular biologists. If you are interested in DNA, an organism which consists of almost nothing else is clearly the right place to start studying its effects. Even without DNA, the phage sat right on the borderline that defined molecular biology before it even had a name: the place where unliving molecules whose properties and interactions could be analysed according to ordinary physical and chemical laws combined to form a living whole whose workings were still mysterious. The summer schools where scientists skilled in other disciplines could learn the care and manipulation of phage were the equivalent of baptism by immersion in the new science.

Edgar was a Canadian then, a young man who had spun like a demagnetised compass needle in his adolescence before suddenly finding his direction as a graduate student in biology at the University of Rochester in New York state. He is an old man now, who has almost entirely renounced science: he lives in the hills above Santa Cruz with his wife, who teaches shamanism in a beautifully spare and elegant wooden house with a teepee in the garden behind for spiritual purposes. In his study are three computers and the usual bewilderment of books. On one wall hangs a rack of wooden Native American flutes.

'I was a very poor student. Looking back on it, I'm quite sure I would have been classified as either attention-deficit disorder or these mild forms of autism – Asperger's syndrome, something like that. I did miserably at school. I flunked out of college in my sophomore year and worked for a year in an oil refinery. At that time, I didn't think I was ever going to go back to college because I'd flunked out, but then I realised that I didn't like working in the oil refinery on swing shifts. It's terrible. So I worked hard enough to go back to college.

'Then I started doing well and I spent a lot of time in the genetics department, doing undergraduate research, things like that. I also spent a lot of time in the literature department. I was writing poetry and stuff and publishing a little magazine. But they still didn't think I could get into graduate school or anything; I didn't have any idea about doing that. So I went away for the summer to a forest insect laboratory, a Canadian government laboratory, for the summer and then I got a call from this genetics professor at McGill [University], saying he'd managed to get me into graduate school. There I was in the middle of Canada; he was calling me from the east coast, this professor.'

Pure luck, and the courage of his professors, had dropped Edgar into the start of a revolution. His professor at Rochester was part of the 'Delbrück clique', the little group of phage specialists who founded molecular biology in the Fifties. Most of them were, like Delbrück and Crick, former physicists. Edgar, as a graduate student, was immediately swept into the excitement of the phage world and soon into the forefront of biology. It was in his lab, when he had one of his own at Caltech, that Richard Feynman spent a summer in the late Fifties learning biology.

Edgar loved the 'Mastermind' aspects of genetics. The phage themselves are too small to be seen by anything except an electron microscope. They are detected because of their effects. A small boy is detected by the disappearance of cake, said Delbrück; phage is detected by the disappearance of infected bacteria. You can actually see this happening, in a test-tube full of cloudy bacterial suspension which suddenly goes clear as they all burst to release their viral load. More often, it is visible in the clear spots or 'plaques' that appear in patches of bacteria as they grow in petri dishes. If each dish represents a different strain or a different question, you can read an array of them almost exactly as if you were playing Mastermind with nature.

The early molecular geneticists prided themselves on this, as on much else. Brenner wrote of their confidence that no one outside their small evangelical sect knew anything worthwhile. Bob Edgar regards it as part of the beauty of genetics that he had no need to understand the chemical details of how genes encoded things. He could discover their inner logic with elegant deductive experiments.

'When I was doing this work, every morning I was eager to get into the lab because I could read the petri dishes after four hours or something. So if you're real eager, you can do two experiments a day. You do an experiment, the phage gives you the answer, you read the petri dishes, design a new experiment, asking the organism a new question, and there's a great exhilaration in that whole process.'

Brenner and Crick, he thinks, were responsible for the most beautiful of all these experiments, the 1959 ones that established that the genetic code came in triplets, read one after the other, whose meaning was dependent on where you started to read from. This meant that a mutation which broke a protein by removing one base pair might have its effect by turning a triplet for a valid amino acid into one that said 'Stop here.' It also meant that two more deletion mutations in the same place might remove the stop triplet altogether, and let normal protein synthesis resume.

This meant sorting through tens of thousands of plates of bacteria with mutated phage on them – phage were essential to this experiment because they have such brief lives and short genomes that there was a reasonable chance that random mutations would knock out three successive base pairs in the course of a couple of months. The reasoning behind it was perfectly clear, and the result, when it came, could be seen at a glance. In fact, they were seen at a glance, when Crick and Leslie Barnett came into the phage lab (set up in the corridor of the annexe to the zoological museum in Cambridge) at ten o'clock one evening, and saw the plaques on their plates of bacteria, where the reconstituted triple mutant phage had been at work.*

The elegance of this experiment fascinated the young Bob Edgar. 'I loved genetics, I loved genetic analysis. I loved

* The story comes from Horace Freeland Judson, *The Eighth Day of Creation*, pp. 465–67.

doing things in an indirect way, getting an answer to something. That experiment to me is the greatest thing that's
ever been. When he gave a talk about it at Caltech, I said,
"Sydney, you're pulling our leg," because he can do that. I
was sure that this was just too bizarre, too incredible.'

Most of the earliest tricks used on the worm had been
developments or adaptations of the methods that had
worked with phage, and with bacteria when phage proved
too simple. That is why Brenner spent a long time working
on mutants which were temperature-sensitive, or would
grow only in the presence of particular nutrients, which
the experimenter could adjust to taste. 'Conditional'
mutants, apparent only under certain conditions, were an
essential tool for the early geneticists. They could grow
their organisms at a low temperature, around 15°C, and
then expose it to high temperatures to see what genes broke
and thus learn what genes it had contained.

The picture of the genome came out over time almost
in the way in which the image on a photographic negative
emerges under the right sort of chemical treatment; and
the whole idea came to Edgar in a sort of vision.

'I was at Cold Spring Harbor and I had been talking with
Alan Campbell, who's a *Lambda* [phage] guy from
Stanford, about these funny mutants we were finding. They
were conditional mutants but we didn't realise it. He called
them "sensitive" mutants and he had found some in *Lambda*.
Dick Epstein had found others and we were trying to make
sense of them. We'd had this conversation and then I said (I
was at the blackboard), "Conditional mutants: it seems like
a general category."

'Then, that evening, we had a party (we had a lot at Cold
Spring Harbor). Lobster, clambake, a lot of beer. Anyway, I
went back, I went up to bed. I'd really had much too much
to drink, I was feeling nauseous. I lay on the bed, it starts

moving around. You figure, "Am I going to get up and get sick, just go and vomit and get rid of it, or just lie here in the bed and roll around like this?" And then I have this vision, see, and I see this genome! I see this genome with all the genes laid out, every single gene lying out there, and this gene is doing this and that gene is doing that and I see this whole picture all coming from the conditionals. It was like everything was all there, it all pre-existed in my head. Then I spent the next ten years, proving it was right.'

By the end of the Sixties, he had managed to find eighty different genes in his chosen phage by using these conditional methods. It was the same sort of effort as Brenner and Muriel Wigby were to put into finding and establishing the five hundred worm mutants they published in 1974. But the effort left him bored and exhausted. 'I'd done as much as I wanted to do. I'd started a whole industry: they had a meeting with thousands of people at it, devoted entirely to the assembly of T4 phage. There were people working on the different genes that make the head, the different genes that make the tail. But once the general principles were laid out, I wasn't really very interested.'

He moved from Caltech to Santa Cruz, and in about 1974 went to Cambridge on a sabbatical to learn about the worm. That made him a disconcertingly old part of the set-up, for the lab was already filling up with young and hungry American post-docs, and some of them were getting ready to return to the States with the worms and the knowledge they would need to start their own labs.

David Hirsh was almost the first man to set up a worm lab in the USA. He is the head of biochemistry at Columbia University now, a neat, tanned man, quick, quiet and precise, like a friendly gecko, but still caught in gusts of enthusiasm when he remembers his time at the LMB in Cambridge. 'It was heaven on earth,' he says. 'Unbridled

energy and high intellectuality and talk all the time about
interesting questions to work on.' Giants walked this earth:
'Max [Perutz] was working in the lab, doing crystallography, while he was the director of the lab. Sydney was there
till three in the morning every night and talking constantly.
Francis [Crick] was going in every which direction but still
interested in coding and decoding suppression but, at the
same time, starting to get interested in neurobiology. It was
really something.'

Hirsh had come there as a post-doc in 1968, to work for
Crick and Brenner on tRNAs, the tiny fragments of specialised RNA which ensure that amino acids are fitted in the
right order to the messenger RNA templates for a protein. He
was looking for one which does its job wrongly, and hooks
an amino acid onto a triplet codon, UGA, which ought to
stop protein synthesis. Since this stop codon itself might be
the result of a mutation which made a defective protein
product, a further mutation in the tRNA would suppress that
mutation and restore normal function. This was typical of
the problems that were left once the grand decipherment of
the code was complete, playing Mastermind at a chemical
level. But after two years there his project was done; and his
funding was for three years. The last thing he wanted to do
was to return early, so he spent some time in the library,
looking for new problems, and Brenner suggested that he
go to work on the worm.

Thirty years later, he's not sure whether the animal itself
was what charmed so many people into working on it so
successfully, or whether it was Brenner's magnetism and
the enormously glamorous future that Brenner predicted
for the project. 'He was able to recruit relatively daring,
extremely well-trained people': the implication is that such
people would have succeeded at anything they did. There
were certainly other primitive animals whose study led to

scientific success: the seaslug, the planarian worm, and even the amoeboid slime mould. But Brenner's enchantment was not cast over a random animal. The great problems that he proposed to examine through the worm had all been chosen before he fixed on his organism, and he had chosen it carefully as being the most illuminating one through which to examine the problems that interested him, and which he could make the smartest young post-docs believe were the most interesting and important of their day. Everyone in the lab was aware of his work on the worm. Hirsh says that 'There was this mystery that surrounded Sydney and Muriel Sharp [sic], alone in this room in the back for all those years, working on this thing, and then there was Sydney sitting in coffee room, talking about how he'd just found *unc*-10 with a passion – and how we were going to map out the whole structure.'

Hirsh recalls wryly that none of the early post-docs achieved much on the worm in their time at Cambridge. But when he returned to the States, to a job in Boulder, Colorado, he took worms with him, and started to use them to look for temperature-sensitive mutants in the early embryos. He was never much interested in the main part of Brenner's programme, which had to do with the generation of behaviour. But the disruption of embryonic development promised to give an insight into the most basic processes of growth and development in the worm. The very first cleavage of an embryo from one fertilised into two cells normally happens quite symmetrically: the egg splits into two equal parts. But there are mutants in which it always splits into one large and one small, like a tiny cottage loaf. Hirsh asked himself what kind of gene could possibly cause that; indeed, what kind of a gene could possibly arrange for the egg to divide as symmetrically as it normally does.

Since a lot of these mutations are lethal when they appear, they were among the things that could best be investigated with temperature-sensitive variations. Obviously, a temperature-sensitive mutation cannot exactly duplicate the ways in which a normally lethal version goes wrong. But the process of genetic investigation is meant to discover everything that's needed for normal things to go right. The temperature-sensitive failures just have to occur in the same genes as fatal failures; they do not have to be the same. The logic of heat-sensitive mutations is delicious. Most of them consist of an amino acid substitution which produces a protein that is almost the right shape but has a weak link in it and is much less resistant to heat than the normal version. So their function as a way to develop a picture latent in the genome arises directly from their influence on the form of the proteins. And it is this kind of deep link between form and function which supplies a lot of the intellectual beauty of molecular biology.

Hirsh would start with a plate of worms, and mutagenise them as usual, to try and establish a group of mutant strains. They would all be grown for a couple of generations – about a month – at 15°C and then their descendants grown at 30°C. The process was easy enough if you were looking for adult mutants. 'If you wanted to look for temperature-sensitive twitchers or rollers, then you would look and see this guy twitching at high temperature and then pick him and see if his offspring did it. But with embryonic lethals, there was no return. They were lethal.' If they showed up, all he would have was a dead embryo in an unhatched egg, which was not something he could breed from.

The best he could do was to look for surviving relatives on a plate where there were lots of dead embryos, and breed from them. Since the dead ones would be homozygous – they would have two copies of the fatal gene – some of their surviving siblings would be heterozygous and have only one copy: that is why they survived. But, because the worms are hermaphrodites which have sex with themselves, the genes would be shuffled in the descendants of these heterozygous hermaphrodites according to the normal Mendelian rules. Half their offspring would be heterozygous and show no sign of damage, though carrying the lethal gene. A quarter would be homozygous for the good form, and neither show damage nor carry the bad gene. And a quarter would die, since they were homozygous for the bad form. It's another example of the enormous importance the worm's sex life had for the project.*

There was a further, and important, complication in that quite often a homozygous mutant, which should have died, could survive as an embryo if its parent made the protein that the embryo could not. If this were something needed for the proper formation of an egg, it didn't matter whether it came from the egg or the surrounding gonad, so long as the egg formed properly. So the researchers would have to wait another generation to see the effects they were interested in, while brooding on the idea of really early environmental influences.

It was this interest in early embryogenesis and its surroundings that led Bob Horvitz to suggest to Hirsh that it should be his lab which studied in cellular detail how the gonad grew once Sulston and Horvitz had shown that lineaging was possible. That was why Judith Kimble ended up doing the gonad's lineage. The possibilities seemed endless in those early days. Hirsh remembers, 'The worm was a new toy, a new gadget. So even if your lab was focused on

---

* Around then, Jonathan Hodgkin was working on his thesis, known around the lab as 'Sex and the Single Worm'.

early embryonic lineage, you couldn't help but overhear in the corner that someone was just *playing* with the way the worm responds to chemicals, or senses heat.

'We all worried about getting tenure. We all worried about getting respect and all those things; so we all had basal levels of ambition, at least, and I suspect we all had a little more than that to keep going the way we did. But the worm just offered you so many chances to play in so many ways and, to some extent, still does.'

In 1975 a community began to form in the diaspora of post-docs from the Brenner project in the LMB. It was not in the least self-conscious at first, and only became so as a result of a collision of effort. For while Hirsh, in Colorado, was working on his temperature-sensitive mutants, Bob Edgar, who had crossed the Atlantic in the other direction, was trying to do temperature-sensitive mutants at the LMB in Cambridge.

His plan was simply to repeat what had worked for the phage T4. 'I thought, first of all, it can't be that hard: after all, I just ran the map into the ground with temperature-sensitive lethals, so why don't I do this with the worm?' But the worm breeds much more slowly than the phage, and has far more genes – we know now that there are about 2,000 with visible effects and 19,500 in all. 'It got to be a huge morass,' says Edgar. 'It became clear to me that it was totally overwhelming, that this was not what I wanted to be doing.'

Then he discovered that Hirsh, back in Colorado, was publishing papers on temperature-sensitive mutants and that they contained all the results he had been working towards. This was not pleasant. He considered taking Hirsh's work further, but the task would have been immense and not necessarily productive. Instead, he started looking at temperature-sensitive mutants which moved oddly: twitchers or rollers, as they were classified. What made this strange and interesting was the way the worm moults. As in all nematodes, its muscles and nerves are hung from a flexible external tube of collagen; and like all nematodes, it rebuilds this external skeleton three times, when it moults. It is a way of life which allows for the extraordinary adaptations parasites can require in different stages of their life style. If a nematode starts life as an egg in sheep faeces, before hatching inside a beetle's lung and then making its home in the intestines of the sheep that eats the beetle along with the grass (this is not a particularly far-fetched example), it needs entirely different exteriors for each stage of its life. Sometimes it wants to be smooth and slippery; sometimes it needs hooks to hang inside an intestine with. All these can be provided by different collagen structures after each moult.

The temperature-sensitive mutants that Edgar now set himself to collecting had slight deformations in the collagen. Instead of assembling into a regular pattern, it formed a spiral, so the animal thrashed as it moved, twisting to either the right or the left, depending on which way the collagen was misfitted. It turned out that there were whole cascades of collagen genes which were switched on at each moult, but the families were subtly different, so that some genes were involved in L1 (the first larval stage) and L3 but not in L2 and L4, and the mutants were therefore perfectly normal for part of their growth. This was fascinating and elegant, but it didn't seem to be going anywhere. What Edgar really did for the worm project was to provide it with the sinews of a community: meetings and a newsletter.

He believed that part of the success of the phage group had come from its feeling of being a community, and encouraged by Brenner, he set about replicating some of this for the worm. 'Sydney brought me into his office and sat me down and told me that he wanted me to be active in nurturing the American worm people: he sort of selected me as the person to do this. I don't remember exactly the things that he said but I took it on as a mission, partly because he wanted me to.

'It became clear from my own work that this whole enterprise was really the same enterprise that we'd started earlier with the phage, only now it had to be done on a large scale and the only way it was going to happen was by people really collaborating.'

With help from Bill Woods, who had been collaborating with Hirsh, he organised the first international worm meeting at Woods Hole in 1975. 'My recollections of it are vague, but I do remember getting up on my hind feet and telling people it was very important to collaborate.'

There were fewer than thirty people at the meeting, and not all of them worked full-time on the worm. John Sulston went – he didn't want to at first, but he enjoyed himself once he did. But to other people, the little huddle of worm scientists were already too tight, self-confident, and convinced of their own destiny. David Hirsh remembers that 'There was this self-consciousness, that's right, that we were doing something special. Bob Edgar promulgated that a lot. There was a little meeting in Woods Hole, early on; Bob was there, I was there and probably twenty people. And there was somebody there, who had never worked on worms and who never did work on worms after that, who told me afterwards they were sort of offended by this chauvinism, shall we say. They were offended. We were going to solve problems that nobody else was able to solve in any other organism, and that person did not believe this and went off and had a perfectly good career in an entirely different area.'

In fact, the worm group did go on to solve problems that no one else could. Their self-confidence, however off-putting, was to a large extent justified. The worm meetings, held every two years, have grown steadily at 50 per cent a year, until the 2001 meeting, held all over the campus at UCLA, had 1,600 participants, parallel sessions, and all the other trappings of a really successful scientific project. Every single one of those participants would, I believe, have told me that the old community was no more. Certainly, everyone I talked to did.

But that was a long way in the future from the first meeting, which led to one of Bob Edgar's most lasting innovations, the *Gazette*. The *Worm Breeders' Gazette* borrowed its title from a then-fashionable flatworm, which was being used to investigate memory. Researchers then believed they could teach a planarian worm to negotiate very simple mazes, and it appeared that the educated flatworm, if ground up and fed to a stupid one, would somehow transmit its knowledge of the maze in the process. This might have led to an educational revolution but the results proved hard to replicate in other animals and even in other laboratories; still, in the springtime of their hopes the planarian researchers had a newsletter called the *Worm-Runners' Digest*.

The first issue of the *Worm Breeders' Gazette,* datelined 'Santa Cruz, December 1975', was a thick stack of xeroxed sheets stapled together, with a hand-drawn cover: the lettering on the title was made from worms, and the border showed geometrical patterns of chromosomes. It contained the kind of information that worm scientists really wanted to know. This had nothing to do with ideals of community. It was remorselessly practical, and does more than anything else to show the conditions under which the first research was done. Almost the first piece of news the *Gazette* brought to the worm breeders' community was that toothpicks were unnecessary. For the first eight years of worm research, all the manipulation had been done with sharpened wooden toothpicks.

'I think one contribution I made was improving the way you pick up the worm,' says Edgar. 'When I went there [to Cambridge], Sydney was teaching us to use wooden sticks. The first thing you did in the morning was you'd take a whole batch of these sticks and for an hour or something you'd be shaving these sticks, getting fine points on them. You have a jar of sticks, now you're ready to work on your worms. Well, these are made of soft wood so as soon as you pick up one or two worms they start swelling, moisture goes into the stick, you can't use it, you have to discard it and take another one.'

So one of the first issues of the *Gazette* contained a short entry: 'Many of us have totally abandoned wooden sticks for transferring worms. Instead a short (1–1½″) piece of platinum wire (32 gauge) is sealed in a Pasteur pipette holder. Platinum is soft and a sharp hook or spatular end can be shaped with a razor blade and forceps. Various "tools" of varied shape can be manufactured. Such implements are readily sterilized in a micro-burner.'

Platinum, being soft and easy to shape, was much better than more obvious forms of wire: using a metal needle was like 'trying to pick up spaghetti with an axe', says Edgar. He is still proud that 'pretty soon, everybody in the lab was using these little devices except for Sydney, who didn't approve of them at all'.

When I went on a worm course at the Sanger Centre, twenty-five years after the *Worm Breeders' Gazette* started publication, in the cool and spacious surroundings of a really modern lab the worms were still being moved with little platinum-tipped picks, flattened and curved at the ends like hockey sticks, and then moistened with sticky bacteria so there was no need to scoop up a worm. You just touched it with the end of the pick.

The first issue of the *Gazette* also had advice on how to send worms through the post: you folded a couple in moist filter paper and then wrapped them in kitchen foil to seal, before popping the whole lot in an envelope.

The other truly memorable piece in that first issue was the sizzling worm counter, a device invented by Sam Ward, a post-doc who went on to have a distinguished career. It seems at first as if this is going to be a properly scientific piece of apparatus. 'In genetic crosses and in experiments measuring the number of progeny produced by a worm, the counting of progeny is tedious. It is often important to kill off or remove all of one generation before they reach maturity in order to prevent a second generation from confusing results. We have found that a miniature soldering iron can be used for killing individual animals on a petri plate. The killing is reliable and fast if the tip is cleaned of roasted worm periodically. If the soldering iron is connected to a counting circuit, such as those used in resistance-type bacterial plaque counters, the counts are recorded automatically.'

The entries in the *Gazette* were short, seldom more than three hundred words long. The print was horrible and the diagrams coarse. But it worked. I don't think there could be a better record of the things the community thought important. It was a printed equivalent of the coffee-room inquisitions at the LMB. When Sydney Brenner appeared, short and terrifying, to demand that you give an account of your work at two minutes' notice, the results should have been something like the reports in the *Gazette*. This was a community built around constant mutual testing. It was a fellowship among steeplechase jockeys. To be a proper part of it they all had to clear the hurdles and go on clearing them. Everyone remembers the friendship and the mutual excitement of the LMB. But they also remember the way in which – even if their boss appeared to take no interest – they had constantly to justify to their peers what they were doing, to explain the reason for their experiments and what they hoped to find by them. It was not enough to be doing something just because it was possible.

Long after Crick and Brenner had left the LMB, their spirit pervaded the Wednesday afternoon divisional meeting. Richard Durbin, now the deputy director of the Sanger Centre, and even then an acknowledged whizz-kid, remembers how frightening it could be in the early Eighties. 'It was in a very small room; everyone sat on the floor and round the chairs; and it was the talk you were most scared of giving. There was a very inquisitorial sense about it, and the question that you got asked, whoever you were, repeatedly and in various forms, was "What is the point of doing this? Why is this the most important thing? What is really worthwhile in this? You shouldn't just be doing something because you can do it easily or because other people do it. What really, in the largest scale you can think about, is this for?"'

Marty Chalfie is at Columbia now, like David Hirsh; but he had been a schoolfriend of Bob Horvitz in the Midwest. Both men moved east for their university education: Chalfie was finishing a doctorate at Harvard medical school when Horvitz came over from Cambridge, England, for the first worm meeting at Woods Hole. Horvitz encouraged Chalfie to get a post-doc at the LMB, where he went to work on the bits of the nervous system that John Sulston was just abandoning for the embryonic lineage. He, too, remembers the curious balance of solitude and mutual support in the labs at the LMB. 'You were on your own. You didn't report to anyone, you didn't answer to anyone, you didn't get a project from somebody to work on.

'That meant that some people got lost but it was very stimulating. I remember that, at one time at coffee hour, somebody came in and said that the astronomy department had just gotten some of the Voyager-Jupiter fly-by films and they were going to be showing them in about half an hour down in the astronomy department, and about twenty-five people got up and we all went down to see these things. There was just an excitement about it.'

Chalfie went on to invent the enormously important and influential technique of GFP tagging, which is a way to mark particular genes in the worm so that the proteins they code for glow green in fluorescent light. This is done by very detailed molecular engineering: the jellyfish fluorescence gene is spliced into the worm's own DNA so that whenever the protein you want is made, so is the visible marker. This was a breakthrough far outside the worm field, and earned him publication in *Science*, one of the two most prestigious journals in the world. That came in 1994, but the tradition of the worm community was still thriving then, for he published the technique five months before the official and prestigious version in *Science*, in a note he put in the *Worm Breeders' Gazette*.

It was not just the self-conscious self-confidence of the phage group that Bob Edgar tried to transmit to the worm community, through the meetings and the *Gazette*. The tradition of sharing results, and techniques, came also from the heritage of genetics. The traditional study of genes ignored their chemistry, which meant that the only way to find out about them was in contrast or combination with other ones. A single version of a gene is almost a contradiction in terms to a geneticist. If it does not have alleles (variations), if there is not another form which can make a difference, it might as well be invisible. So publication and sharing are essential to making progress for geneticists. Biochemists, by contrast, had a tradition of owning the molecules they worked on. These attitudes were to contribute something to the later clashes over patenting DNA. Patenting, or even owning, a molecule was not a thought which would ever occur to a geneticist, or even to most molecular biologists.

Most of the humour in the *Gazette* came from the covers, which grew steadily more extravagant. My favourite, and perhaps the most famous, came on the front of volume 8, number 2. It shows *Caenorhabditis brobdignagiae*, a giant worm, peering down a microscope at a petri dish full of terrified humans. There was almost always a worm on the cover: the one showing a 'temperature-sensitive mutant' had a very long worm, wearing, at various stages down her body, sunglasses, ski-ing clothes, a bikini and a heavy shawl. But there were sly jokes slid into the text as well: Phil Anderson – a man gigantic in height, girth and self-confidence – reported his method for growing really huge quantities of worms in a 300-litre barrel, ending up with a pellet of hard-packed centrifuged worms which weighed 1.35kg before being broken into nuggets 'which look like garbanzo beans'. Almost all of them were dropped into liquid nitrogen and frozen; the report concluded that worms treated in this way 'taste like chicken'.*

The *Gazette* was virtually the only written source of worm information for a decade. Bob Edgar and Bob Horvitz went out as missionaries to Japan, to teach courses in the worm; and anyone trained in an existing worm lab would learn the basic techniques needed. The *Gazette* is full

---

* Leon Avery, who now maintains the *Worm Breeders' Gazette* online, is supposed once to have spread some on a hamburger bun and eaten them.

of what John Sulston calls the 'artisan' part of science: in
one issue, Sam Ward confronted the problem that arose
when worms which had been sorted into little pits like fox-
holes escaped into the wrong ones, and so got classified
wrongly. 'Do your worms crawl out of their microwells
polluting the genotypes of their neighbours? We find that
2ml disposable plastic beakers that are sold for use with
Technicon Autoanalyzers make excellent tiny petri plates
(they cost .07¢ apiece). They stand in the 24 well Falcon
microtitre plates and no worm yet studied has made it from
one well to the next. They can be sterilised with uv . . .'

The manipulation and care of worms sounds ridiculously
easy. In fact, even the most basic skills are extraordinarily
difficult. In an effort to find out how the worm community
recruits new members, I sat in on a course in autumn 2001
at the Sanger Centre. It was taught by Jonathan Hodgkin,
who had done his PhD on the worm under Sydney Brenner
in the early Seventies, and ended up as a professor of genetics at Oxford, and as the curator of the genome: the man
who has to approve the names of every gene anyone has
found. So he is a gatekeeper as well as a founder of the
community.

Away from the agar, he is slightly ungainly, like a heron
walking on dry land, or a man whose spectacles are too
heavy. But when he sat at a microscope, all that disappeared, and he was as poised and economical as a heron
fishing. He was completely still except for the hand that
moved the worm pick. He didn't even seem to breathe. It
was almost like seeing a watchmaker at work. All around
him were smart and ambitious young graduate students,
people who would consider themselves up to almost any
intellectual challenge, and certainly able to pick up a worm
from a plastic dish. Yet many of them found it extraordinarily difficult to do so, and difficult even to recognise the

one mutated worm they should pick as it crawled, or failed to crawl, among a hundred others beneath their microscopes. It may be that this dexterity and keen observational skill are two reasons why almost every successful worm scientist turns out to have had a childhood which involved fiddling with gadgets. There could be few better preparations for sorting through a dish of worms with a pick than soldering on a crowded circuit board – even when the implement you're using is not in fact a modified soldering iron.

Hodgkin inserted no grand gadget. But he did discover a trick that stopped them wriggling. 'It is convenient to be able to pick up many worms on one worm pick, especially when picking males for many parallel crosses. I have found chilling worms to be helpful. Place the plate to be picked from on ice for 20–30 minutes. The worms become completely torpid. Scoop up a sticky gob of bacteria on the end of a platinum wire worm pick, and pick up the torpid worms by adherence to this gob. When the mass of worms and bacteria is placed in contact with agar at room temperature, the worms instantly recover their mobility and swim off, having suffered no ill effects. It is possible to pick up and transfer as many as 100 wild-type males at once by this means.'*

Donna Albertson compares some aspects of science to cooking: you need luck and skill to follow the difficult recipes. In fact, some textbooks are remarkably like uninspired cookbooks, with detailed recipes for every procedure; and a scientific paper usually has a note at the end on the recipes ('Protocols') and a list of the ingredients used to produce the results. But science is more like the restaurant trade than domestic cooking. It is constantly

* *Worm Breeders' Gazette*, 8/1: 52.

looking for novelty; and the most valuable recipes have not yet made it into the textbooks. Albertson was one of the first American post-docs, converted to the worm by Sam Ward when he left the MRC and taught her at Harvard. She was another tinkerer by upbringing: her father was a former navy engineer, who later got a job converting early cinemas to play talkies. She came for a summer to the MRC in 1974, just before Bob Horvitz. She was interested in the serial slicing of the nervous system, and started working with John White, to whom she was later married for a while.

Some of the other tricks in the early editions of the *Worm Breeders' Gazette* were really complicated. John Sulston and Donna Albertson found a way to open the tiny eggs from inside the worm for staining: they froze them between two glass coverslips, which were normally placed on top of microscope slides, and when they pulled the coverslips apart, half the egg stuck to each, leaving a speck which could easily be stained to bring out its inner structure.

All these little hacks, spread through the *Gazette,* must have helped to keep the community together as the diaspora grew. The *Gazette* itself grew fatter as the number of worm labs, and worm researchers, grew. The covers grew funnier and more elaborate. But it was growing beyond the point where everything could be an in-joke. One kind of tombstone for the original community came in 1986, with the publication of the first textbook on the worm. *The Nematode Caenorhabditis elegans*, 667 pages long, covered almost everything then known about the worm, and had a foreword by Sydney Brenner. All of it had been discovered since 1966. It was a triumph, of course, but it also marked the point at which worm science had become so well established that it was no longer attractive to the kind of people who had been drawn to the animal at the beginning. Like molecular biology itself, the study of the worm had begun

in a place where there could be no textbooks, and some
people missed that.

At around the same time Bob Edgar retired, sick at heart
with his profession. 'Science has become more and more
ordinary. I mean, it's so dreary. "OK, here's a gene. You go
and you sequence and then, once you've got the sequences,
go and look at which cells the genes are active in, where the
RNA is, blah-blah ... You've got all these different techniques; everything's doable. But it's just all hard work now.
[With phage] I could ask a really interesting scientific question and have an answer the next day. Now it takes you a
team of labourers working for four years to get something.'

Edgar is remembered with affectionate bemusement by
his former colleagues. He obviously came to seem at times
like an Ancient Mariner, trying to turn them into a reincar-
nation of the phage group, and talking about a community
spirit which none of them trusted quite as much as he did.
Perhaps it was the kind of community you could not wholly
feel at home in, because it had no place for the old. There
is a rule at the worm meetings that no one, however distinguished, or however important their discovery, may speak
for longer than ten minutes. The only person for whom
this was ever waived was Sydney Brenner. But Bob Edgar's
abnormal story is a mutation which reveals something
important about the phenotype of normal worm scientists.
Only in someone for whom the passion of scientific discovery has suddenly died can you see how great it must be in
everyone else and what a huge hole would be left in their
lives if it were gone.

There is also a genuine difference of style between the
phage group and the worm researchers. The first generation
of worm researchers took pure genetics about as far as it
could go. There came a point where, if they were really
going to understand a gene, they had to understand the

chemicals it was made from and the reactions that these took part in as they went about their genetic business. It was no longer enough to discover, and then to list, all the genes that provide ways in which a worm can break. It was necessary to discover exactly what mechanisms they broke; and how they contributed to the normal functioning of the animal when they were not broken. That was the final stage, of Brenner's original plan. It was to involve explorations of unfathomable complexity.

8

The DNA Revolution

Around 1980, the worm disappeared. This makes it difficult
to write about the next part of the story. Until the late
Seventies, the investigation of the worm had a visible reality. You could have watched Nichol Thomson slicing them,
or Sydney Brenner picking through his mutants at two in
the morning, and got some sense of what they hoped to
find. Physical experimentation with the worms continued
throughout the Eighties and Nineties and led to some
important discoveries (which will be dealt with later).
Indeed, people still slice worms, and watch them under
microscopes, and they still learn things that way.

But for about fifteen years the most original and important bits of the effort to understand the worm concentrated
on its genetic material; and at the scale on which it is interesting DNA is so small that it can't be seen, can't be smelled,
and doesn't feel like anything. It turns into a molecule made
up of large and simple ideas crammed into a very small and
complicated space. This makes it hard to visualise what the
biologists are doing when they manipulate it; because of
that, it is hard to understand what tremendous difficulties

they face. If you really want to get a feel for the ways in which it fits together and can be manipulated, you need pictures. Better than that, you need models, in two or three dimensions. There are, in fact, undergraduate courses in molecular biology where the students cut out bits of shaped paper, like jigsaw pieces, representing the different pieces of the cell's mechanisms, and push them around until they can see and feel how the messenger RNA forms from the DNA template, and then is chopped and rearranged until it has become a template for a gene, before the transfer RNAs are hooked on in order to make a protein at the ribosomes. It sounds confusing. It *is* extremely complicated. But seen as a physical model it all makes sense.

The difficulty of understanding arises partly because the scale on which everything happens is far outside the range of our normal experience. There is a thought experiment that shows this clearly enough. Imagine a piece of string as long as your arm: about a metre. Now make it so thin that it is invisible, however closely you look; that is just about imaginable. If the gossamer strand were crumpled into a ball, it could still be rubbed between your fingers and felt. So you are not imagining it nearly thin enough. Make it thinner yet: so thin that it can't be seen under a microscope but still forms a single strand; that is about the stage where it is hard really to imagine such a thing. But if it were rolled up into a ball, it would still be just about visible with a powerful microscope, so it is still too thick to be DNA. The real measure of the size of the molecule that all this fuss is about comes when you realise that there is a metre of it in every one of the trillions of cells in your body, and it is so thin that the whole string, the length of your arm, is coiled up into a space so small it's invisible.

When people read about the sequencing of a genome, they believe that it is rather like playing a film through a cinema projector: you just feed a long ribbon through the right reading device, and then its meaning is projected brightly for everyone to see. It isn't. It couldn't be. The illusion of a projector or a tape head which can read the DNA off a chromosome is something produced by the metaphors we use to describe it. And you cannot understand what happened next with the worm without realising just how much effort and artifice went into the reconstruction of its genome.

Alan Coulson grew up in Cambridge: when he heard about the structure of DNA in a science lesson in 1960, he thought of it as a local story. He also understood at once that it was wonderful. 'That's one of the few things I remember from school lessons. I always liked biology but I couldn't honestly say that I had a burning desire to do this. In fact, I wanted to do psychology at one time. My A levels weren't good enough so . . .'

He emerged from Leicester Polytechnic in 1967 with an HND in applied biology, 'which was actually of no relevance whatever to what I've done since then'. He saw an ad in the *New Scientist* for a post as Fred Sanger's assistant at the LMB, and got the job: a rangy, quiet, bearded man, who at first appears even quieter and more withdrawn than John Sulston, but, while Sulston will eventually start bubbling, Coulson remains focused and dry. His is almost the ideal temperament for the patient, exhausting labour of drilling down through the squishy chemical complications of a chromosome all wrapped in its proteins until you get to the hard-edged digital certainties of the DNA sequence. There is probably no one alive who knows more about how hard this has been. In the forty years of his career in science, he has been at the forefront of every attempt to get at the sequence of DNA and RNA, and has put his patient determination up against every discouragement.

The first difficulty is that it is all the same. Not only is it unfathomably thin: it's almost unfathomably monotonous. Every one of the 3,000 million bases in a human can be only C, A, T or G. The joke bears repeating: trying to find the sequence you want really is like trying to find a haystraw in a haystack. But there is a further difficulty: the DNA comes in long, continuous chromosomes, as if your haystack were made from six immensely long plaited haystraws. They needed to break these plaits into manageable stalks, which proved very hard to do.

When Coulson came to work for Fred Sanger in 1967, people had known for fourteen years that DNA carried the molecular code, but only a few minute and unusual stretches of it had been deciphered or sequenced.

All the worm's geneticists in the early part of the project had been using mutations to break the worm's DNA into its separate genes, or units of function. That was rather like trying to understand a language by splitting a sentence up into separate words and, by looking at them in different contexts, discover what each word does for the sense of the sentence. But there are other ways in which you might want to break up a sentence or a genome. You might want to split at random, between letters, if you were trying to destroy the sense.

It can be done chemically, and some enzymes which will do it to RNA had been discovered and purified by the Sixties; but they did not work on DNA, and using them was very slow. It took two years in the early Sixties to isolate, purify and then sequence a stretch of RNA seventy base pairs long. At that rate, it would have taken 3 million years to sequence the worm, and 90 million years for the human genome.

The first real progress was made by subverting the mechanisms that nature has evolved for breaking the DNA in cells. Bacteria have very complicated methods of preserving the sense of their own DNA: these are necessary if they are to reproduce themselves faithfully. But it is just as essential for them to be able to destroy the sense of viral DNA. Otherwise, the phage virus (which is nothing but a minimally wrapped streak of DNA) will take over the cell, and the bacterium, instead of making the proteins it needs to thrive and grow, will make only fresh copies of the phage, until it bursts and dies.

In the early Seventies, it was discovered that bacteria manufactured enzymes which destroyed DNA in exactly the way needed to make phage harmless, by chopping it in places which destroyed the sense of a gene by fastening onto particular sequences of bases and breaking the chain there, just as you could destroy most of the sense of a sentence by removing the letter pairs 'en' or 'st' wherever they appeared.*

These were more useful, though more complicated, than the similar enzymes known for RNA, because the RNA ones worked at a single base pair, and not at a pattern or string of bases. So they were only useful on very small fragments of RNA indeed. Large ones would be chopped into bits far too small to be of any use to analyse the string they had come from. It would be like trying to reconstruct a sentence from a scrabble bag of the individual letters that it used.

The 'restriction enzymes'† that work on DNA are not

---

* deroy mo of the sse of a ce by breaking it.
† So named because they restrict lengths of DNA. Scientific language is for the most part remorselessly bathetic. Everything is named after its function, which is why there are things wandering around inside your cells called spliceosomes. One of the few exceptions to this are names invented by Brenner, among them 'Amber codons', originally a pun on the name of its discoverer, a Dr Bernstein, whose name is the German for amber.

destructive by accident. Obviously, it would be more useful to have enzymes which broke the DNA at gene boundaries, like a program which would split and sort the words of a book at the spaces and punctuation marks. But such enzymes don't exist. If DNA is the book of life, as we are told, Nature uses a typewriter with no spacebar. That is one reason why, before the human genome was completely sequenced, it was supposed to contain 80,000 genes: two years later it was supposed to contain only 30,000; and now there are people (Brenner among them) who think the right figure should be closer to 60,000.

Enzymes which break strings of DNA at gene boundaries would be so useful that someone would have built them if it were possible. But it is not, and the reasons are built into the genetic code. This code gives the rules by which a sequence of three DNA letters specifies a particular amino acid. These triplet sequences do not overlap: they are read one after the other. The proof that the code must have this character was one of the first things Brenner had worked on in the Fifties. But it has the awful consequence that you can't in principle look at a piece of DNA and read the meaning off without knowing where you should start, and that information is not always obvious or even possible to decipher. All* genes start with the sequence ATG, but this sequence can occur without signalling the start of a gene at all. If you have chopped from the middle of a larger sequence a string of DNA which runs ATGCATGCGTGGG, it is impossible to tell from first principles whether this sequence ought to be read as part of a gene starting ATG CAT GCG TGG G, or as part of a different gene which starts a little further on,

---

* (protein-coding bits of) – a detail which doesn't affect the main argument.

A TGC ATG CGT GGG,* a reading which changes the place of two of the four original amino acids, and transforms the other two into new ones.

This theoretical difficulty was shown to be practical as well from the moment when, in 1977, Fred Sanger sequenced the first whole bacteriophage,† and found that the 5,386 bases coded for ten genes, and four of these overlapped, meaning that within those genes the same string of DNA was being used to specify two different proteins, depending on the context in which it was read. There is one place in the sequence where all three genes overlap, so that the cell reads three different protein sequences from the same piece of DNA, depending on which gene is being used.

These complications are not just practically important, though obviously they are. They should also help to destroy the idea that the sequence itself contains all the information about an organism, even one so simple as a phage. The information in the sequence is absolutely necessary to understand an organism. But it is not sufficient without all kinds of extra-sequential information, starting with the knowledge of what counts as a gene and when.

Early sequencing had to confront the problem that the steps from phage to bacterium, and then from bacterium to worm, were huge. It was very much quicker, as well as more thorough, than the purely genetic methods used before (Bob Edgar had taken ten years to find and isolate eighty genes in one phage in the Sixties). But it was still

---

* At the same time, it is also possible that the other, complementary strand of DNA is coding for a completely different bit of protein, when read in the opposite direction.
† OX 174.

extremely slow if you were interested in organisms larger and more complicated than the phage.

By the late Seventies, Coulson had helped to sequence the bacteriophage *Lambda* (which is a lot smaller than Bob Edgar's phage, T4). It took three years with a side-track into human mitochondrial DNA. Elsewhere at the MRC, Bart Barrell was working on the other sequences of direct interest to humans: cytomegalovirus, a usually harmless condition related to herpes, and the virus for Epstein-Barr disease. These were about a quarter of a million bases long, so they were far larger and more complicated than the viruses that prey on bacteria.

To go from there to the genome of a multicellular animal did not look at all practical. When Sanger accepted his second Nobel prize in 1980, he reckoned sequencing at a rate of 500 to 1,000 base pairs a day was fast, and that was, for the time, a wildly optimistic estimate. Bart Barrell took five years over cytomegalovirus, finishing in 1989, at an average speed of around 1,000 base pairs a week.

Almost the first work that Sulston and Brenner had done on the worm had been to work out how much DNA it had: they came up with a figure of about 80 million base pairs, about 1,600 times as much as *Lambda* phage. This meant that when Sulston finished his description of the embryo, a complete sequence of the worm was a ludicrously ambitious project. What could be done, he thought, was to make a map of manageably large fragments of the worm's genome which would put them all in order. As more and more genes were discovered, it was obviously going to be necessary to find where on the chromosome they were located, as a preliminary to sequencing them. What this came down to, in practice, was discovering which fragment of worm DNA, represented in thousands of copies in a library of modified bacteria, was the one that had any particular gene on it. If you were going to do that, it was necessary to place all the fragments in order. This was jigsaw puzzling on a truly heroic scale.

While he was pondering this problem, in 1983, Fred Sanger retired, and Brenner suggested that Coulson move over to work with John Sulston. Though the two men had been working in the same MRC building for fifteen years, they had never had a real conversation. They were on different floors, and the three different floors of the MRC regarded one another with a certain suspicion: each thought they were the only ones who had got the balance between reductionism and really interesting biology exactly right. Coulson knew vaguely about the worm, because there were lecture programmes in which the different parts of the MRC talked to one another. But he had never seen one; and in fact a knowledge of worms down a microscope was entirely irrelevant to the work he would have to do. He went to Sulston's office to find out what he was up to: 'I just remember going to see him and he sort of explained his thoughts on this and then suggested we go and continue the conversation in the pub, which seemed a good idea. So that was what we did and that's how it started, really.'

Going down to the pub was not merely sociable. In the cramped conditions of the LMB, it was almost essential if you were going to talk. Sulston's 'office' at that time consisted of a metre of bench space, with another metre of desk sticking out from it at right angles, for a telephone. That may slightly overstate his space: I believe he graduated to a telephone only after he had started the mapping project. On the other hand, he was thirty years a scientist before he had to write a single grant proposal, and the lack of a desk meant also a lack of bureaucratic obstacles between him and the intrinsic difficulties of his task. It should also be said that space was the only resource in limited supply at the

LMB. There was never any trouble getting equipment, and American scientists, remembering the lack of bureaucracy, are still delighted by the profusion of the stockroom. They would give you anything you wanted, if you could find the space to put it. Marty Chalfie remembers wondering whether they had a ruby laser, which he needed for a particular experiment. When he went down to inquire, all that he was asked at the stockroom window was 'Which wavelength?' – a reply so generous that for a moment he didn't understand it.

Coulson and Sulston meshed almost at once. There is a determined quiet practicality about them which reminds me of Kipling's engineers. Even today, Coulson does not seem to think of himself as the sort of biologist much interested in animals. 'I wouldn't classify myself as a worm biologist at all. If John had happened to have been working on *Drosophila* or something, I'd probably have been working on the *Drosophila* genome. So I'm not a worm biologist; but I do feel a kind of loyalty to the organism. That's probably a stupid thing to say but I do think it's a great organism. It does have – the community has – attributes which were missing from the *Drosophila* community and to their detriment because they had all kinds of competing groups who were doing different but uncooperative things.'

The techniques and tricks needed to get enough DNA for their analysis were pretty well understood by the time Coulson started work. Worms in bulk are measured as if they were liquid, by the millilitre or even larger chunks. To make a sample of worm DNA, Coulson started with a millilitre of worms, and ground them up in liquid nitrogen: when that evaporated, he was left with a powder of deep-frozen worm tissue, which was mixed with detergent and salt to crack the cell walls and then a couple of other solvents until pure alcohol extracted the DNA itself. The extreme cold of the sample preserved the DNA from chemical attack by the enzymes in the broken-down cells, and in the end he was left with 'a big goop' of the stuff. That was given a brief treatment with restriction enzymes, and the resulting fragments – still looking like a single goop – were placed on plastic covered in a clear gel.

By pulsing an electric current through the gel, he could measure the length of the fragments. The shorter they were, the further the current would move them. And these methods of 'gel electrophoresis' had become so sensitive by the early Seventies that they could distinguish two lengths of DNA of which one was only a single base pair longer than the other.

That kind of precision wasn't immediately necessary for mapping. All he wanted for mapping were the bits about 40kb* long. The restriction enzymes had given them ragged ends which exactly matched those of fragments of a bacteriophage which had been chopped open with the same enzyme. This virus normally keeps its DNA in a doughnut-shaped ring. When the rings closed again, many had a length of worm DNA inside them. There was one further elegant and necessary subtlety: they also carried a natural bacterial gene for antibiotic resistance.

Now his goop contained a few thousand hybrid doughnuts of worm and virus DNA: cosmids. The goop was mixed with another, containing phage proteins which let it slip through the outer membranes of bacterial cells, and the new, infectious artificial viruses were poured into a flask of bacteria. After a short incubation, tiny drops of the new mixture were distributed on dishes of antibiotic jelly. This would kill all the bacteria except those which had taken up the cosmids, and thus acquired antibiotic resistance along with worm DNA.

* A thousand base pairs is a kilobase, or kb; a million is a megabase, or mb.

So any spot of gel with flourishing bacteria on it would be the beginning of a library, where the inserted doughnuts of DNA worked sufficiently like a virus to be copied each time the bacteria multiplied; but what was being copied was in large part a fragment of worm DNA. After a couple of days, there were hundreds of thousands of copies of worm DNA, providing plenty of material for analysis.*

They called them libraries: once their cosmids had been grown in innumerable copies inside bacteria, some reference copies were frozen down in glycerol and from then on were to all intents and purposes immortal. The tubes of frozen bacteria could contain DNA from anything at all. The combination of techniques was what led Sydney Brenner to exclaim that the whole of biology had been transformed in this way, which meant we could now put DNA from a duck and from an orange together and be absolutely certain that they would combine into a nutritious hybrid. I once asked Coulson to show me a bit of a library, and he came up with a sheet of grey plastic about the size of a postcard, neatly gridded with black spots which we supposed for a moment were worm; until he realised it was actually John Sulston's DNA.

If the library's fairly long stretches of DNA were extracted from their cosmids and treated for longer with more restriction enzymes, they broke up again. When all the different bits were spread out along gels, they formed a pattern which with the eye of faith looked like a barcode. To the naked eye, it looks more like gridlock in a rush-hour of slugs. But the essential point is that these patterns recorded

---

* This technique is called cloning, which is confusing if the word makes you think of Dolly the sheep. But it has nothing to do with the techniques used to manufacture her, except that it involves DNA.

the distances along the sections of DNA between the fixed patterns that restriction enzymes fasten onto, and these distances form distinguishable patterns which can be used to identify – with any luck – a stretch of cosmid. This technique, known as 'fingerprinting',* seems to be another thing that originated in Brenner's torrent of ideas, though it was actually developed by another MRC person, Jon Karn. Coulson managed to get the technique working on a much larger scale.

If they found the same pattern in two cosmids, they knew they overlapped. What they did not know was where and how, or in which direction along the chromosome, the cosmids did their overlapping. The fingerprinting process destroyed all the information about where on a cosmid the fingerprinting pattern might lie, and in fact the sections that made up a fingerprint might be scattered all over the cosmid. So two cosmids which shared a fingerprint might coincide only at their tips, or they might contain almost identical portions of the genome.

There's no theoretical limit to the technique: by breaking the worm's DNA into pieces like this, and then multiplying each fragment in bacteria, it ought to be possible to get the whole of the worm's genome reproduced in manageable fragments which could themselves be broken up again and sequenced in due course.

The task Sulston set himself was to get all these portions in order: to come up with a map which would show where and on which chromosome each cosmid had come from. It still wouldn't be a complete sequence of the genome: that

---

* It is not the same as, though related to, the 'genetic fingerprinting' used to identify people, which involves completely sequencing known fragments of DNA which vary widely but unimportantly between people.

was to be nearly a decade of work away. But it would be an enormous amount of work.

The first problem is simply the number of clones required to make a library of the worm. In the early Eighties, the only thing in which DNA could be grown was a bacterium.* But bacteria are small, and the largest fragment of worm DNA they could be induced to grow was only about 40,000 base pairs long. You would need 2,500 of those to cover the whole genome, even if they fitted together closely without any overlapping. But the whole point of the technique lay in the fact that the cosmids they used overlapped. So it was necessary to start with something like 17,000 cosmids.

Of course, to sort through thousands and thousands of cosmids looking for overlaps was a job beyond even the most patient geneticist. Traditional genetics is like a code-guessing game: there is some clear intellectual satisfaction beyond the routine. But sorting and shuffling through the endless barcode patterns is more like playing a game of pelmanism with a pack of 17,000 cards. Sorting through half a million worms looking for mutants can be done reasonably quickly by eye: a good geneticist can count a worm a second, or 3,600 an hour. In 1978, Jonathan Hodgkin had screened 40,000 mutants looking for sex-determining genes. But genomes are far too big for that to be done by hand. People were trying, though. 'With this sort of procedure, people were doing it in a quite manual way. They'd get the half-dozen clones and put them each in a tube and—' Here Sulston, normally a precise and articulate speaker, gave a distant note of remembered anguish, like a wolf trapped in a cold store.

---

* This meant that E. coli was being used both to feed the worm and to grow fragments of its DNA; later, it would be used to knock out particular genes in the worm.

Sulston remembers that this slow, sequential mapping was 'anathema: the bane of people's lives'. The problem was that two clones might overlap for 98 per cent of their lengths, and there was no way to tell this. All that the tests would show was that the two had some sequence in common. But they couldn't show where on the cosmids the shared sequences were. Sulston decided that the answer lay in computers. Instead of having graduate students all over the world toiling each at their own fragments of the worm genome, he would look at hundreds a week, and later hundreds a day, looking for fracture patterns or 'barcodes' which could be recorded and fitted together in a computer. The plan seemed absurdly ambitious, and it certainly required skills entirely different from those Sulston had honed in his study of embryological development.

The worm was no longer ignored by people who worked on other organisms, but it was certainly patronised by the fruit-fly people. Alan Coulson remembers that a drosophilist told Sulston he was wasting his time, since the *Drosophila* community would know everything in about five years.

The rivalry with *Drosophila* cropped up in several people's reminiscences. Phil Anderson, at the University of Madison, said, 'From the mid-Seventies to the mid-Nineties, *C. elegans* was often in a beauty competition with *Drosophila*. I've been on review panels, looking at grant applications, where a *C. elegans* grant would come up and there'd be disparaging remarks, often from the *Drosophila* community, about "Well, this is so limited in its scope and abilities and the methodologies are not there. It'd be better to do this in a fruit fly or in something else," and it was very vulnerable in those times.'

It was the development of the map that finally established *C. elegans* as a force to be reckoned with, according to Anderson; but for about two years progress was very slow: preparing the libraries was a skill which had to be learned, and working with the results was even harder. The first way they had to digitise their sample involved tapping each band on the gel with a stylus connected to the computer, which recorded their positions. It was boring, error-prone and bad for the wrists. Even worse was the way they recorded the overlaps (or 'contigs') the computer had found. Coulson simply marked them as pencil lines in notebooks, rubbing them out and redrawing when a contig was extended by fresh discoveries. They are still in his office, with the numbering neatly written in beside them: 199 overlapping with 223, and perhaps with C30 as well. But C3099 makes an even bigger contig – the thought of flicking through thousands of pages of such notebooks was giddying.

By 1984 they could announce their technique in the *Worm Breeders' Gazette*,* but progress was slow. They believed they had mapped about two fifths of the genome, but it was broken into 325 separate contigs, and they were analysing only about 100 clones a week.

At this point, Sulston, like Brenner before him, plunged into computers for the answer. He learned Fortran, a language notably less ghastly than the assembler Brenner had used, though still a long way from fun; and spent about two years working on a group of programs which could digitise photographs of the 700 or so gels containing 17,000 fingerprints, and then scan them for similarities and classify the results. The original plan to scan the whole nematode's nervous system into a computer had failed in the early Seventies because of the difficulties of recognising distorted slices of worm. The X-ray films of slightly

* Vol. 8, p. 2.

radioactive gel fragments that Sulston was working with were much more irregular than the slices produced by a microtome. But computer technology had advanced a great deal, too. The workshop at the LMB built a film scanner to replace the scientist with a stylus, and Sulston managed to write the drivers that would let the scanner recognise and compensate for the irregular squishiness of the gels it was studying.

The same improvements in computer power finally allowed John White to complete his plan of a four-dimensional microscope, one that could record and store a complete record of change in a three-dimensional organism. Part of the software for that was written by Richard Durbin, yet another promising young scientist who found himself spending more time with computers. But Durbin's career – he is now the deputy director of the Sanger Centre – marks a watershed, because he came along at exactly the moment when the development of computer programs was the technique by which molecular biology could be most advanced. He helped Sulston with the final step of his suite of programs: a replacement for the pencil lines in a paper notebook.

Durbin came up with a quicker and more effective way to choose the best overlaps among a set of clones which had been recognised as matching by the scanner software. Sulston wrote a pretty front-end* which enabled him to check whether the scanner had correctly recognised the bands in the gel it was examining, and to correct its mistakes if it had made any. It was simple, sturdy and efficient.

Coulson still uses Sulston and Durbin's program for fitting fingerprints together. 'I'm probably the only person

* The bit of a program you can actually see, and push a mouse or cursor around.

who does but I'm so used to it – it actually has features which I find really convenient which the newer generation of mapping programmes . . . for me, they have too many unnecessary buttons. I find I can do everything I want with the old programme.' He is not, I think, an enthusiast of computers: on the shelf above the workstation in his office that runs Durbin's later and more elegant software, AceDB,* which can display everything that's known about the worm's genome, there is a bright yellow copy of *Unix for Dummies*. But the mapping project marked the point at which computers became indispensable for molecular biology.

Once the software was done, by the middle of the Eighties, the mapping project began to be obviously useful to everyone in the field. An important part of this success was that it was open and collaborative from the beginning, but the collaboration was a means to more effective competition in science, rather than an attempt to eliminate it altogether.

The relationship between competition and collaboration in the worm project was always complicated, and it became immensely important when Sulston and Bob Waterston moved on to head the publicly funded human genome project. The ethos of the LMB, and of Cambridge science in general was not in the least opposed to competition. The ten Nobel prizes won by the lab since 1945 were not obtained by coming second in the chase for important results. Sulston himself, once he had decided to do the embryonic lineage, steamed ahead of the Göttingen team after some hesitation but quite without remorse. In his later battles against Craig Venter's attempt to get a private sequence of the human genome, he was quite exceptionally

* Written with Jean Thierry Mieg: see Chapter 10.

bloody-minded, and determined. At one stage he broke a hip in a motorcycle accident and discharged himself from hospital five days later, into the care of Bob Waterston, who had never practised as a doctor after qualifying, so that he could attend a crucial funding meeting in London in his wheelchair.

But side by side with this toughness went an extreme reluctance to compete where it was not necessary. The spirit of the coffee mornings was an ineradicable part of the LMB. People shared every idea they had. Often large changes in careers came about after someone was shoved into an office doing something unexpected because there was no room for them elsewhere. No one huddled over his work. In part this generosity derived from the kind of self-confidence that almost transcends arrogance. The people who flourished at the LMB expected to be excellent at anything to which they turned their hands. Brenner, entering university at fourteen, was of course exceptional by any standards, but most of the early worm scientists had a remarkable range as well as number of degrees: Waterston started as an engineer, then became a medical doctor, and ended up as a research scientist; Sulston's career ran from chemist to geneticist to the pure observational biology of his lineage tracing work; Bob Horvitz was a computer programmer and mathematician as well as a cell biologist. These people were in favour of free trade in ideas, partly because they knew they would come out towards the front of any competition demanding intellectual competence and hard work. But they also shared a very strong streak of idealism. They believed that the unfettered exercise of their talents benefited the whole of humanity. The differences between art and science are considerable, and I don't want to blur them, but it seems clear that both artists and scientists are borne up by a feeling that the work that is most important to them is also vital to the world. This may not be rational, but it is a load-bearing part of the soul's constitution.

Besides, the more you share your results, the more likely it is that other people will find something interesting in them and come to work alongside you. So from the very beginning the mentions of the project in the *Worm Breeders' Gazette* were concerned to show that researchers would not lose by sending their clones to the project in Cambridge: 'Although the existence and position of your clone become public, the clone itself (and any contiguous clones) do not. Enquiries about the region immediately adjacent to your clone are invariably directed back to you before we send out any material', as Coulson explained to the community.*

The genetic map was a perfect example of a public good within the worm community. Everyone benefited from the fact that no one owned it. In fact, the balance of advantage is even sharper: everyone benefited more from public ownership than they would have done from owning all the data themselves. 'It is clear from the outset that the physical map will only become a reality as a communal project', wrote Coulson, Sulston and Brenner in the *Worm Breeders' Gazette*, announcing their project. 'Numerous markers will be required to align it with the genetic map.' From the beginning, they wanted people to send them any fragments of cloned DNA which had known genes on them, to be fingerprinted and made part of the map. 'In return, we can send you any flanking cosmids that we find, as well as any others that you need.'*

One of the things that had attracted Sulston to the mapping project, when he at last looked up from his embryos at the worm world around him, was that it was obvious there

---

* *Worm Breeders' Gazette*, 10:2.

was beginning to be serious competition among the different worm laboratories. The second wave of efforts to clone genes were very competitive: they had to be, when the whole structure of science rewards the person who finds something first, and the time needed to find and clone a single gene was two or three years – about as long as, a century ago, it would have taken to equip and mount an expedition to find and map the South Pole. It is quite possible the money and manpower were comparable, too, and with stakes that high it was obvious that the field would get crowded: 'people were beating each other up to be the first to publish,' says Sulston now. No one wanted to be remembered as the Robert Falcon Scott of a project like that. Mapping the entire genome was an entirely characteristic way round the competition, by finding something so hard, and requiring so much hard work, that no one else would be tempted to try it, just as the embryonic lineage had been. It is ironic that it led him into the largest and best-funded competition in the history of science, one which more resembled the space race of the Fifties and Sixties than anything which had taken place in laboratories before.

* *Worm Breeders' Gazette*, 8:2.

9

The Sequence

The worm need not have been the animal whose genome led directly to the human's. That it became so was almost entirely a result of the hard work and collaboration between Sulston, Coulson and Bob Waterston, for there were many other organisms, most of them more famous or more popular, whose DNA was being mapped in the Eighties. Mapping and ultimately sequencing were the obvious next step for any creature which had been intensively studied. The *Drosophila* effort has already been mentioned. But *Drosophila* is a much more complicated animal than the worm, and it has a lot more DNA (though this turns out to contain rather fewer genes). The natural early target for sequencing efforts was brewer's yeast, *Saccharomyces cerevisiae*, which has only one cell, but is a proper eukaryotic animal with its chromosomes wrapped up in a nucleus and not a bacterium, whose DNA is loose in the cell. The man with big plans for *S. cerevisiae* was Maynard Olson, who worked just down the corridor from Bob Waterston in St Louis.

Olson had planned to make a physical map of *S. cerevisiae* as early as 1979, when he arrived at Washington

University in St Louis. He needed to make only about
3,000 cosmids to do so,* but even that was a huge number
to manufacture and sort through by hand. Besides, the
work was incredibly monotonous, or so it seemed to
Waterston at the time. All the techniques for extracting
DNA and growing it had to be worked out by hand, and then
each and every one of the clones had to be made and
processed. It was completely opposed to the traditions of
biology until then. Waterston remembers a sign above the
door of one of the Woods Hole Labs: 'Do every experiment an even number of times, preferably less than three.'
Doing the same thing two or three thousand times over, in
order to get results which looked exactly the same – for
they could not be properly analysed for a couple of years –
seemed to take tedium into a new dimension. But the
point, Olson used to say, was that the new experiment was
his own. 'Maynard is very fond of saying that when he
watches what most biologists do, he can't think of anything more tedious because they come in and they basically
do variations on somebody else's experiment,' said
Waterston. He started to talk about the role of a tolerance
for boredom in scientific success. 'Doing worm genetics is
pretty monotonous. You have to sit and you look at a plate
for funny-looking animals and then you start doing process
and you have to set up the process and you have to score,
you have to pick . . . Repetition is a key element. But I
think the point is that biologists look down on it and are
afraid of it.'

Waterston was still working on muscle proteins in the
early Eighties, but he knew a lot about Maynard Olson's
work, and he knew, of course, that John Sulston was plan-
ning to map the worm. In 1985 he got a sabbatical to return

* Bob Waterston, interview with the author.

to the MRC to work on muscle proteins. But – in keeping with the generally cramped tradition of the place – there was no room for him in the muscle lab. Instead, he found himself in the same room as Coulson and Sulston, as they tried to get all the worm's DNA to grow inside bacteria. Quite large chunks of it wouldn't. By 1986 they had managed to organise their 17,000 cosmids into 700 groups, known as contigs because the segments within them were contiguous. But the gaps between the contigs remained hard to fill. The underlying problem was that certain portions of worm DNA simply would not grow well in bacteria: *E. coli* did not absorb them, or did not grow if it had absorbed them.

Waterston had brought with him some tricks from St Louis. 'I was primed for all this because of sitting next to Maynard and being interested in this. It didn't take a genius to see that if you really wanted to get all those fifty muscle genes, a map was going to be the fastest way to get these. So I sat there and I brought some of the technology that Maynard had been working on in his lab, because, by that time, already it was clear that the cosmids that were being used were not going to close the map.'

When he returned to St Louis, however, he found there had been a revolutionary discovery in Olson's lab. They had found a way to get yeast cells to treat inserted DNA as if it were a whole chromosome, and so grow chunks almost ten times longer than could be grown in bacteria. This meant, of course, that it would be very much easier to close the gaps in the worm sequence by growing and fingerprinting much larger fragments of DNA inside yeast cells. It had a further advantage: yeast would grow some short fragments of worm DNA that *E. coli* wouldn't touch. These YACS*

---

* Yeast artificial chromosomes.

proved essential when working with any eukaryotic genome, from worms to humans.

Waterston started to work with both Maynard Olson, next door to him, and Sulston and Coulson across the Atlantic. He spent a couple of years making a library of YACs for the worm. It turned out that they could not be finger-printed in the simple way that had been used for the bacterial cosmids, so he would send his YACs to Cambridge and the Cambridge team would send their cosmids to him. Instead of fingerprinting the cosmids and the YACs would be hybridised to each other by seeing if strands from yeast and bacterial clones which had been separated by heating would combine with each other when they cooled down. If they did, the two clones obviously came from the same part of the genome. This technique was astonishingly successful. A Japanese molecular biologist, Yuji Kohara, had come to Cambridge for a sabbatical to learn about the worm that year (1986) and set to work matching all the worm YACs with the bacterial cos-mids. Within seven months, they had joined up half of the gaps remaining, and within a couple of years the map was almost entirely complete: by 1992, the mapped and joined data covered more than 90,000,000 base pairs of the worm's genome, and there were only six small gaps left.

The map had become really useful years before it was complete. What made it so was the way it combined with all the other knowledge accumulating about the worm. It cannot be too often said, in all the hype about genomes, that the knowledge of a sequence on its own is almost use-less. Sulston says, 'It's not until you put markers onto the physical map that it becomes interesting. That's what made it work. It was not a theoretical concept, it was a pragmatic way of moving forward.'

The map offered a new dimension to all the other knowl-edge. Without the map the library was like the web used to

be before search engines, a territory not only unexplored
but almost impossible to explore. The technique for finding
and identifying a gene then was almost incredibly laborious.

The way that worked best was to use 'transposons', little
genes which tend to hop around the chromosomes when it
is copied to make eggs or sperm. Quite often they pop up in
the middle of other genes, disrupting their function. One of
them is common in some strains of worm, which have
around fifty copies. The transposons' fingerprinting pattern is quite easy to recognise. So what people did was look
for mutations which broke the gene they wanted in a worm
strain known to be full of transposons, grind up the mutant
worms and make a library from their DNA, then sort through
the 17,000 clones of this new library looking for transposons where they shouldn't be; and hoping, when one
was found, that it had popped up in the middle of the gene
whose mutation they had looked for. It was perhaps slightly
more efficient than blasting a shotgun into a reservoir and
hoping to hit a passing trout.

Getting the physical map changed all that for ever. Most
known genes had already been placed on a genetic map,
which measured how likely they were to be inherited inde-
pendently of other mutations, and thus roughly how close
they were on a chromosome. This meant that, every time
one of those genes was found in a contig on the physical
map, the scientists knew which cosmids to look in for its
neighbours on the genetic map.

Just as the embryonic lineage, the detailed anatomical
mapping of John White's slicings, and the catalogue of
mutants had formed a tripod on which reliable knowledge
could be based, which relied on all parts of it, the map
became part of a new tripod. The other two legs of the new
tripod were the earlier catalogue of genes which had been
identified by their effects, and the increasing knowledge of

the precise sequences of some of the genes so identified. When these three new sorts of knowledge were combined, it was possible to know of some genes exactly where they were found on their chromosome, what was their sequence, what protein it specified, and where in the worm it worked.

If that could be known for every gene in the worm, the resulting knowledge would be pretty much what Brenner had originally dreamed of. But each step of this under-standing was harder than the last and needed more heroic calculation. The allocation of genes identified by their effects to particular 'linkage groups' or chromosomes had taken about ten years, and required no computers. That gave a rough measure of the physical position of the genes on the chromosome, though what it actually measured was their likelihood of being separated when new copies of a chromosome were generated in sperm or eggs. The physical map took another ten years, and required quite a number of computers, though it still made no real use of the embry-onic Internet. The sequence took a further ten years; and the final part of the process, understanding which proteins the genes make, and what these actually do in the worm, could take until the end of this century, and require the kind of computer power that is hard to imagine.

Both Sulston and Waterston published their map data pretty much as they got it, and by 1989 it was available from their computers. But the process was so slow and error-prone – it could take a month for a long e-mail to ric-ochet from Cambridge to St Louis through the system known as Bitnet – that at first they posted each other tapes every couple of months, rather than trying to e-mail such huge files.

Still, in 1989, they announced in the *Worm Breeders' Gazette* that there was a 1,200-baud modem in Cambridge through which the database could be accessed, and two –

twice as fast – in the States, one in St Louis, and one at MIT. They had also by then got a clearer grasp of the size of the genome they were going to have to deal with: the original calculation of 80mb had been made by John Sulston in the very early Seventies by a comparison with *E. coli*, which was then believed to have a 4mb genome. But Yuji Kohara had gone back and discovered that *E. coli* had a genome closer to 5 million base pairs than four, so the estimate for the worm was revised upwards to 95mb. By that time Sulston was determined to sequence the whole of *C. elegans*, not merely to map it, and he began to try to raise funds for this. It was, at the start, the need to raise funds for the worm which led him to the human genome project.

The idea of sequencing large genomes down to their smallest constituents seems to have first emerged seriously in 1985. A European proposal to sequence the whole of *E. coli* had been floated in 1980, but it was then technically impossible. Since the worm has a genome twenty times as large as *E. coli*'s, the technical impossibility went in spades for that, although it is possible that Maynard Olson and John Sulston had been thinking about it quietly. But the plan might never have coalesced out of the cloudy brilliance of late-night boozy talk – Waterston remembers a dinner party at John Sulston's – had it not been for the determination of the University of California at Santa Cruz to spend a really humungous sum of money so that people would take it seriously.

Some parts of California exude dynamism and excitement. Los Angeles and San Francisco have world-class art museums. Santa Cruz has a museum of surfing instead. It also has three large bookshops in the main street downtown, some excellent wineries in the hills around, and a university which does not believe in exams and so had a serious image problem by the early Eighties. Bob Edgar, then chairman of the biology department, says, 'People think Santa Cruz is just too beautiful to have a university: that everybody'll be surfing and they won't, you know . . . It's not a good place to go if you want academic respectability.'

The new chancellor, Robert Sinsheimer, was determined to change that. Sinsheimer had himself been a distinguished molecular biologist: his lab at Caltech had purified and genetically mapped the phage *phi*-X-174 which Sanger later sequenced. But when he went to UCSC in 1977, he found – according to the university's official record – that 'The outside world (as well as segments of the Santa Cruz community) had come, however wrongly, to view UCSC as a flaky, hippie school, with a questionable academic reputation. Vietnam war demonstrations, drugs, and the campus's counterculture increasingly strained town–gown relations and UCSC's reputation throughout the state. Enrolment figures were down and there were rumors (unfounded) that the campus would be closed for budgetary reasons.'*

Obviously, one way to restore the reputation of the university would be to spend a great deal of money. The first project he tried was a telescope:† the cost of upgrading the famous observatory at Mount Palomar with the biggest telescope in the world was put at $500 million in 1984. But most of that cost would come from building to very fine tolerances the huge mirror that such a telescope would need. If smaller and lighter mirrors were used, the cost could be dramatically reduced: thirty-six hexagonal mirrors, synchronised by a computer, would cost an eighth as much, and that was a sum which Sinsheimer almost raised. He

---

* The quote comes from the official history at http://library.ucsc.edu/reg-hist/sinsheimer.html.
† Source for most of this is Cook-Deegan, *The Gene Wars,* p. 81ff.

persuaded a Mrs Hoffman, the widow of a Volkswagen concessionnaire, to leave the university $36 million for the new telescope. She signed the papers the day before she died. It was to be the Hoffman telescope, a name which would guarantee her and her husband a sort of immortality.

But the money was not enough. The telescope, if it were built, would cost nearly as much again, and Sinsheimer found it very hard to raise the extra money, since none of the new donors could hope to have their names on it. When he appealed to his old workplace, Caltech, for help, he got more than he could handle. Another charitable foundation, built not on Volkswagens but on Superior Oil, offered to fund the whole thing if the name were changed to the Keck Telescope, and that is what was eventually built, on Mauna Kea volcano in Hawaii. The Hoffman foundation was not going to give all that money to something called 'Keck', so Sinsheimer had to hand back their cheque.

He tried again. If the biggest telescope in the world were not to come to Santa Cruz, perhaps he could make the university the home of the most powerful microscope ever imagined: an institute which would unravel all human DNA. This was a project infinitely more ambitious (and, it turned out, thirty times as expensive) as merely building the biggest telescope in the world. He asked Bob Edgar and a couple of other biologists on the staff to arrange a meeting where the feasibility could be discussed. Fred Sanger, now retired (Bart Barrell was his successor), sent a note of encouragement.

Edgar's contribution, he remembers, was to ensure that some of the people who came had real, practical knowledge of what was being done and what was possible.

'I wanted to invite John Sulston because I knew he was already working on the worm. Wally Gilbert, I think, was the only one that we invited who was sort of one of the Grand Old Men. I studiously did not invite Jim Watson, because he would just dominate the whole thing and wouldn't contribute to the discussion. So there was a mix of older people and young people who were really working, like John.'

But he went to some lengths to track down Gilbert, who was travelling in the South Pacific between jobs. After his Nobel prize in 1980, he had left Harvard to run a biotechnology firm he had helped to found. It was not a success. In the two years he was CEO it lost around $25 million, and he resigned in December 1984 and returned to Harvard, where he became chairman of the department of biology. The meeting converted him to the idea of sequencing the human genome, at least in principle. Edgar says, 'The older people were surprised at how far things had gone, when they heard John talk about what he was doing and how he was going about it. And so, pretty soon, everybody realised, "Wow, this is going to be done pretty soon and the question is how."'

Edgar and his colleagues did not really believe, when they arranged the meeting, that the human genome could be sequenced – and with the technology available in 1985, it certainly could not have been. But they did think it might be mapped, as the worm's genome was being, and that the mapping would be useful in itself. 'The ordered library of cloned DNA that must be produced to allow the genome to be sequenced will itself be of great value to all human genetics researchers.'

By the end of the meeting, the experts had concluded that the full sequence was, for the moment, impossible. In fact, the rate of technological advance they were betting on when they made this prediction was far too fast. They had hoped it would be possible to sequence 50,000 base pairs a week by 1990; in fact, that year most labs were hard pressed to manage 1,000. But even with their optimistic assumptions the task looked impossible. The experts in Santa Cruz decided that what could and should be attempted instead was the development of machines and techniques which would make a map of the human genome thinkable – remember that it is thirty times as large as the worm's, and, since most of it is repetitive 'junk', much more than thirty times as hard to map. But they did believe that it would be worthwhile to sequence in detail a few of the interesting bits of the genome.

To get at this limited knowledge was possible only because of techniques developed in the early Seventies, first by Fred Sanger and Alan Coulson, and then, as they refined their methods, by an independent effort in America: Wally Gilbert at Harvard, with whom Bob Horvitz had studied, found his in the late Seventies, spurred on by the fact that it had taken him two whole years to get the sequence of twenty base pairs using earlier, indirect techniques. Working with his technician, Allan Maxam, Gilbert found chemicals which allowed strands of DNA to be treated so that one of the four bases – C, A, G or T – was rendered fragile so the strand could be later be broken wherever that base appeared when it was mixed with the right solvent.

At the same time, at the MRC, Fred Sanger and Alan Coulson were approaching the prize from the opposite direction. Sanger's method of sequencing involved growing DNA, rather than chopping it up. Using a natural enzyme which copies DNA, he would set out to copy the strand he wanted to analyse. But he did this in solutions in which some of the bases needed to make a new chain of DNA were replaced by subtly mis-shapen ones, which lacked the little stub to which the next link in the chain would normally be attached. This meant that whenever one of these mis-shapen 'dideoxy' bases was used the strand stopped growing at that point.

Either way, you ended up with test tubes containing slightly radioactive DNA strands of different lengths. Somewhere among them would be some broken at every single base in the stretch you wanted to analyse. By running them on four lanes of a gel – one for each base – and then X-raying the results, you could work out the length of every chain. If parallel lines were laid across the gel, with intervals corresponding to one base pair's length, there would be only one lane with a spot in it between each pair of lines. So, by reading off the spots, you could see the position of every base in the sequence; which was what they had been trying to find all along. Gilbert and Sanger got a share of the Nobel prize (Sanger's second), and Maxam got his name first on the Maxam–Gilbert sequencing technique, which is not a bad shot at scientific immortality.

The Santa Cruz campus did not benefit from this in the end. Sinsheimer could no longer reach the Hoffman Foundation directly once it had been brought into the orbit of the University of California's central bureaucracy, and they never managed to get those millions back. He had supposed that the foundation could have built an institute for the study of the human genome for a mere $25 million, which would cost them $5 million a year to run. It never happened. What was left were three things. A dream, a flavour of enormous hype – Sinsheimer had said that the sequence would be 'the basis for all human biology and medicine of the future'; Gilbert called it 'the grail of human genetics [in which] all possible information about the human structure is revealed (but not understood)' – and Sulston's increasing belief that it would soon be practical as well as desirable to have a go at sequencing the whole worm. He would have been happy to try to start mapping the human even then.

However, the ripples from the Santa Cruz meeting spread outwards, into politics. One reason was that the rhetoric was combined with huge sums of money. Gilbert thought the human genome might be sequenced for a dollar a base, or $3,000 million. It soon proved that Sinsheimer's instinct had been right. If you were prepared to spend that much money, it was obvious even to politicians that your science must be tremendously important: there was much more resistance to the idea of sequencing big genomes among biologists than among politicians. Two separate arms of the federal government in Washington became interested: the Department of Energy and the National Institutes for Health (NIH). Naturally they started to compete to fund and control what would happen, and their rivalry drove the project forward for a while. Businessmen were interested, too. In spring 1987, Gilbert tried to form a company, Genome Corporation, which would map the whole of the human genome and supply privately, and for profit, the services that Sulston and Waterston were supplying publicly and freely to the worm researchers.

Despite this obvious difference, both projects were built on the belief that there were natural economies of scale in any mapping project, and that it was wasteful and stupid to compete when everyone would benefit from specialisation. Gilbert's argument was that 'Twenty years ago, every graduate student working in DNA had to learn to purify restriction enzymes. By 1976, no graduate student knew how to purify restriction enzymes; they purchased them, Historically, if you were a chemist, you blew your own glassware. Today, people simply buy plastic.'*

But the stock market had its own reasons for things, and the crash of October 1987 destroyed the hopes of the

* From *Science* 250, quoted in Cook–Deegan, p. 89

Genome Corporation. Gilbert did not give up the idea, and in 1990 he set out to sequence the smallest known free-living organism, *Mycoplasma capricolum*, a bacterium which lives in goats; in the end, this did not happen, and a relative of the goat bacterium that prefers human genitalia was sequenced instead. Even if his efforts did not come to anything directly, they persuaded other people, among them Jim Watson, that there would have to be a serious effort to sequence the human genome. All through 1987, Watson, who was then running the Cold Spring Harbor labs, edged closer to the nascent human genome project. He alone had the reputation, and the ruthless determination to do whatever he thought right, that would be needed to get large sums of public money committed to the project. He acted with a sleepwalker's certainty. All that year he argued that the project could only succeed if it were led by a scientist of international stature. At a meeting in Washington to discuss the genome project,* Watson turned to Sulston in front of everyone and asked, 'Doesn't one person really have to finish up that last 10 per cent and live or die for the thing?' Sulston thought that would give too much power to whoever was chosen. Watson replied that someone had to do the job. Sulston made the obvious retort, that it looked as if Watson were promoting his own candidature, which idea seemed to strike Watson as a novelty. None the less, he got the job, and in 1988 was appointed the first head of the NIH's Office of Genome Research; in 1989, it became the National Center for Human Genome Research, with an annual budget of roughly $60 million.

* OTA, summer 1987. The story is in both Cook–Deegan and *The Common Thread*, by Sulston and Georgina Ferry, an account of the human genome project told through Sulston's eyes.

Sydney Brenner, by then director of the LMB, determined that Britain should not be left out. He did not, then or later, believe that sequencing the whole human genome was real science. He thought it was a kind of glorified stamp collecting: the collection of data for its own sake, with no proper end in mind. What he wanted was an approach much more like Craig Venter's: a determined effort to sequence those parts of the genome which looked as if they contained genes, and not just long strings of repetitive 'junk' DNA which may have no function, and whose function is certainly not understood properly yet. Of course, the function and even the identity of most human genes are still not properly understood, and will take decades at least to unravel. But Brenner believed in going for the parts of the genome that would yield a reward most quickly. He put some of his own money into a project that started at the LMB and then, after a personal letter to Margaret Thatcher, he got some government money for the human mapping project, some of which was done at his own spin-off lab, which he suggested be called 'Sydneyland'.

In all this, you might have expected Sulston and Waterston to move towards the forefront of the effort. In fact, they were preoccupied with the worm. 1989 was the year when their mapping project became a triumph visible to all. Though they had been publishing the data regularly to anyone who wanted to dial in, the real impact was made when Coulson printed the map on sheets of A4 paper, glued them together, and stuck them up on the wall at the biennial worm meeting at Cold Spring Harbor. The six chromosomes, with almost all known genes marked on them, took up most of the back wall of the lecture theatre. Everyone at the meeting was using data from the map by then; and Sulston made sure that Watson came to the meeting and saw the map.

It was a characteristic way for Sulston to make a point, being neither ostentatious nor possible to dispute. The point of the demonstration lay in the fact that there was at the time no comparable map for *Drosophila*. It is true that more individual fly genes had been sequenced, but the existence of Sulston's physical map, and the way it was linked to the genetic map, made it obvious that sequencing anything of interest on the worm would be a matter of routine from then on. The drosophilists would have to sequence every gene from scratch: anyone interested in a worm gene would know exactly where it was in the library, and could very simply order the bacterial clones containing the piece of DNA they were interested in investigating further. All that was left, really, to complete the understanding of the worm's genome, was to sequence the whole lot of it.

The difference between mapping and sequencing is analogous to the difference between two ways of copying a picture. Mapping a genome is like making a pen-and-ink sketch of the original. All the information that seems essential is preserved, but no one looking at the sketch alone could reconstruct the details or the colours that were left out. Sequencing, on the other hand, is the equivalent of making the most detailed colour reproduction possible, from which the original could be reconstructed with almost perfect accuracy. It's obviously far more work. Instead of simply breaking up the clones into enough pieces to give a recognisable pattern, or fingerprint, for each one, the full sequence involves breaking all of them at every single base pair and mapping the results – 80 million times for the worm; 3,000 million times for a human being.

Very large areas of the genome are extremely difficult to map because they are full of repetition or – worse – they contain palindromic sequences where a single strand of DNA

consists of two consecutive sequences which pair with each
other so that it easily folds over and binds to itself in a
double-stranded hairpin which can't be teased out for
analysis. What makes the areas hard to map makes them
even harder to sequence. There are more of them in the
human than in the worm: the amount of 'junk' DNA often
rises with the complexity of the organism, though the relationship is not reliable. Mice have slightly less DNA than
humans; some plants have phenomenal amounts (the trum-
pet lily has thirty times as much as a human); and one
species of amoeba has a genome nearly two hundred times
the size of a human's.

So the practical problems in the way of any attempt to
sequence a complete animal's genome were clearly going to
be considerable. Watson was determined to start on a trial
organism before moving on to the human, and all the evidence suggests that he had marked down Sulston and
Waterston as the people to do it. He did not try to persuade
them directly. What he did was to tell Bob Horvitz, his old
student, who then ran a large lab at MIT, that he thought
Drosophila should be the model organism used to make a
test run for sequencing the human genome. 'He wasn't convinced that the worm community had pulled it together
enough to launch the operation in the way that would really
get the sequence done' said Horvitz later.* That is why
Horvitz, urged on by Sulston, made certain Watson came to
the worm meeting to see how large and how useful the
map of the worm was already by 1989. After he had seen
their work, they ambushed him in his office: Bob Horvitz,
Bob Waterston, Alan Coulson and Sulston went to see
Watson to demand that the worm be funded as the first
trial sequencing project.

* Quoted in *The Common Thread*, p. 67.

The key, they thought, would be to industrialise the process. So far as possible, the work would be done by machines, which would be tended by relatively unskilled labour. The clerical work of calculating would all be done by computers. Computer-controlled machinery would also do all the work of mixing and incubating and reacting that he and Coulson had once done by hand; and it would do so with more and more samples at once. Sulston's confidence was absolute that these things could be done. When Watson asked him how much money it would take to sequence the worm, he replied, in the tradition of the MRC, that if he just gave the worm group $100 million they could do the job in ten years. 'That's not the way we do things in this country,' replied Watson, with so little outrage that he might have made the same calculation himself. But he did agree to give them $3 million for the first 3 million bases, which were to be done in three years, half by Waterston in St Louis, and half by Sanger and Coulson at the LMB, if the Wellcome Trust would chip in another million pounds. Sulston expected that Coulson, who was the man who knew about sequencing, would become the official leader of the British project, if it grew large enough to need one – before then it had simply been the two of them, working so closely together that a secretary in St Louis thought she had been asked to send an e-mail to 'John N. Allen' because 'John'n'Alan' had become a single word. But the principle by which he had lived his life – that the most technically competent person should always run the project – would not apply where such huge sums of money were involved and where managerial responsibilities were so great.

Sulston set out to raise funds from the MRC in Britain. It was the first time in twenty years that he had had to write a grant application, but the forces of bureaucracy had not really closed in on him properly, since the reply to his request came on a single sheet of hand-written fax. Yes, he could have a million pounds with which to sequence the worm for three years. I think he is to this day more proud and gratified by the informality of the reply than by the sum he raised. It showed he was working with the right sort of people. 'There was paperwork that followed it up: we did put in a formal proposal. But it was very brief. I remember Bob Horvitz and Bob Waterston thought it was an absolute hoot, the small amount of stuff I had to write to get that money.'

The other step he took to prepare himself for the task was to learn sequencing. If he was going to direct a sequencing project, he would obviously have to be able to do the work. 'The only way that you can justify having Chiefs and Indians is if the Chiefs can do everything the Indians can, and more.' It is curious that someone so uncomfortable about inequality should end up running so huge a project. Sulston must recognise that some people are more gifted than others. But he doesn't like to do so, preferring to account for other people's success by the virtues of hard work and humility, rather than the sort of apparently effortless brilliance displayed by Sydney Brenner. Of course, a very great deal of hard work goes into being as effortlessly brilliant as Brenner. There is no doubt that when he ran a lab Brenner could have done almost everything that any other member of that lab did, and all his workers knew it. Even in the earliest days of the worm, he did a lot of the unimaginably tedious work of sorting mutants using sharpened toothpicks. No piece of computer software was too boring or insignificant for him to try to write. He learned all the techniques he needed to.

But like many scientists he seems never to have doubted that he could run the rest of the world better if he could be bothered to learn the skills involved in running other people. Sulston, by contrast, seems reluctant to admit that there are any skills involved in running other people. That is not arrogance. It is more a belief that human nature shouldn't be like that, and people should not be so manipulable as we are.

It was certainly in keeping with the anti-bureaucratic ethos of the MRC, in which all management decisions ought to be taken by a working scientist in a corridor while he was on his way between labs or coffee sessions where the real work was done. I cannot help feeling that it expresses something fundamental about the anthropology of science. Hardly any of the people I talked to seemed to have had childhoods. There were exceptions, most obviously Sulston and Brenner himself. But questions about things that happened away from the lab bench seemed to disconcert most people. It was as if there were a moral imperative involved: that nothing outside the lab ought to count. They reacted as if there were something indecent in asking about the passions that had led them into science. It seems taboo to admit that these passions are passionate; yet they must be, when you consider how hard everyone works, how much they risk if they fail, and how much they must give up to succeed.

Even at moments of triumph there is little exuberance shown. A perfect example is Fred Sanger's second Nobel lecture, given when he won a share of the prize for the second time, in 1980. It is difficult to imagine a better occasion for fireworks which would illuminate the enormous future his discoveries had opened up. The ability directly to read, and then to manipulate, the workings of DNA in a cell changed our relationship with the universe around us almost as much as the use of fire, and not just because it is Promethean. It allows the world to be first investigated, then understood, and finally changed in ways that were

simply unimaginable before. Richard Durbin compares it to
the discovery of chemistry in the nineteenth century, which
does not sound very exciting until you realise how much of
our world is made of synthetic substances, and how much
of our food is grown with artificial fertiliser. A world without plastic is hard enough to imagine; a world without
artificial fertiliser (or modern explosives) would be utterly
transformed. So Sanger's Nobel speech might have been an
occasion for looking at the changes in our place in the universe.

The other prize winners took the opportunity. Wally
Gilbert and Paul Berg, who shared the prize with Sanger,
both delivered fairly technical lectures but they kept an eye
on the wider picture. If you want to understand in detail
how Alan Coulson could use fragments of viruses to make
his cosmid libraries, Paul Berg's lecture is an excellent place
to start; after all, he invented the technique. But of course
most people's interest in these techniques is entirely different. We don't want to know how to do it, we want to know
what it can do for us. At the end of the lecture, Berg raised
his eyes to the audience and talked a little about the significance of his work; and Gilbert, the man who later called
the Human Genome Sequence 'the Holy Grail of Biology',
is no stranger to hype. In the latter part of his lecture, he
talked quite a lot about the evolutionary logic that might
underlie the fact that eukaryotic genes are broken up into
exons, separated by non-coding introns, terms which he
himself invented.

Fred Sanger's talk, by contrast, never deviated from
chemistry. He spoke in the robotic passive preferred for
scientific papers. 'In spite of the important role played by
DNA sequences in living matter, it is only relatively recently
that general methods for their determination have been
developed', he started; and he finished abruptly fifteen

pages later: 'The compact structure of the mammalian mito-chondrial genome is in marked contrast to that of yeast, which is about five times as large and yet codes for only about the same number of proteins and RNAs. The genes are separated by long AT-rich stretches of DNA with no obvious biological function. There are also insertion sequences within some of the genes, whereas these appear to be absent from mammalian mitochondrial DNA.'

The lecture told you everything about the difficulties he had overcome, and nothing about why they had been worth overcoming, as though that, to the outside world the most important question, was not admissible as a problem. It is as if Gutenberg had been given the prize for inventing the printing press, and had devoted his entire speech to a dis-cussion of the melting point of lead, the difficulties of avoiding the appearance of bubbles in freshly minted type, and the qualities of wood-pulp, without one word about what might be done with these letters, or the uses of books.

There is something almost Tolstoyan about this humble, patient, undiscourageable stubbornness. It may be signifi-cant that Sanger was brought up as a Quaker, and remained a pacifist throughout the Second World War.* The contrast with Brenner's overwhelming sparkly brilliance could hardly be greater; but the worm sequencing required both qualities.

Sulston and Waterston set off on a tour of the world in 1989 to discover what machines were available to help them in the task they had taken on. They did not believe the machines would be necessary. Sulston later wrote that they could have done the whole thing by hand if they had had to, and that the only really decisive breakthrough in the

---

* He abandoned his early beliefs, but not his ethics, as soon as he discovered scientific research.

field was Fred Sanger's technique of sequencing. Once that had been done, everything else was incremental progress: nice, but not essential.

Richard Durbin believes the sequencing project was really dependent on computers rather than automation. Getting the gels and reading them could have been done by humans, even if it had taken a century. But getting the cosmids into order, and making the map from fingerprints, would have been impossible without computers; and it would also have been impossible to store the data in a way that made it possible to get any useful information out of it. To illustrate this point, consider that half the people alive in the world today have never made a telephone call. If everyone who has ever made a call has their own phone number, then there are about as many phone numbers in the world today as there are base pairs in human DNA. Now imagine all these names and numbers written out in a library full of directories, without being sorted or indexed in any way at all. It would be impossible ever to look up a phone number, even if you had all the phone directories in the world. That is what a gene sequence would be like without computers to organise and make accessible the information that has been extracted from the gels.

None the less, two firms had been building machines that would automate the work of sequencing. Their chief technical innovation was that they labelled the ends of the DNA strands they were sequencing with fluorescent dyes, rather than with the radioactivity that Sanger, Sulston and Coulson had used; and continuously detected the fluorescent fragments as they reached the end of the gel. So there was no need for the X-ray pictures that Sulston scanned in. This was certainly quicker, and might be more accurate than the old methods. Later machines got rid of the gels altogether, and drew the fluorescent tags up hair-fine glass tubes, in which they could be detected and counted by lasers.

The business of making these machines promised to be immensely profitable: ABI Biosystems, a company spun off from the Caltech lab of Lee Hood, one of the pioneers of molecular biology, made a very great deal of money from them, and hoped to make more by controlling the data format used by its machines. The ABI sequencers were enthusiastically recommended to Sulston by a brain researcher in Maryland named Craig Venter, who had been hugely frustrated by the time it took him to isolate some human brain proteins by the old method. 'It was not his lab that convinced us; we did not really see what was going on,' says Sulston,* carefully establishing that he never even then believed Venter's claims without correlating evidence.

What sold him was a visit to Baylor College of Medicine in Houston, Texas, and a trip to Heidelberg, where another machine, marketed by a Swedish company, had been designed. They rewrote their grant proposals to get one machine of each kind, and set them to work in the traditional LMB style. 'We stacked the sequencing machines vertically and we each had a metre or two of desk space. My entire office space consisted of another metre of table sticking out sideways with a phone on it. More like a bedsit than open plan, but a great way to work.'†

Almost the first thing they had to do was to hack their new machines, breaking them up and reassembling them almost as if they were worms. Waterston managed to get the ones in his lab doing two runs a day, doubling their productivity, which had been thought impossible. Sulston

---

* *The Common Thread*, p. 71.
† Sulston, ibid., p. 73

hacked open the proprietary file format in which ABI stored its results. This meant that they could feed the data obtained by the machines straight into their own databases. Before then, everything had come out in the form of print-outs, which had to be kept in ring-binder files – almost a return to the days when Sulston and Coulson did their work in pencil. This was a real disadvantage because the machine's conclusions had to be checked. The worm project employed as few skilled people as it could on sequencing. If the immense amounts of data were to be processed in time, it would have to be done by a process more like a car factory, using robots and semi-skilled labour, than like traditional science. However, the work would have to be overseen by real scientists, who could do the final adjustments that make everything useful, and they needed access to the raw data in order to be sure that the machines had interpreted it right.

Once the file format was cracked, other people joined in and within a couple of months the ABI machines were integrated into the smooth workings of the lab. This speeded up the genome project immensely, and would probably now be illegal under recent legislation designed to protect the profits of intellectual property owners.

The free release of information practised by Sulston and Waterston was not the only model, even among scientists. The yeast workers proposed to get their genome done in the way the frontiers had been settled: the post-docs who were given a piece of the genome to research would retain some rights in it afterwards, as if they had staked a land claim. Perhaps a more Nineties analogy would be to say that they were offered genome options instead of stock options. Waterston explains, 'They could say, "I've got this DNA, I've got all the genes here and I can make a living off of that." And, of course, it doesn't make sense biologically, because the genes that you want to study for any one system are all over the place and they're not really tied together. But the idea was that it would be a way of inducing people to come and work on it because they could have a career pattern.'

Sequencing machines sidetracked the problem of attracting smart and ambitious people to do the drudge work that the yeast scheme was meant to compensate them for. They made possible the free release of sequencing information as it appeared; and that turned out to be part of a virtuous circle. According to Waterston, 'The more we put it out there, the less problem it was; [whereas] the more restricted it was, the more people felt like they had to bargain with us or that they could get special access and they could get a leg-up on somebody. If you just put it out there, then people all had the same footing and they all were then free to talk about it because it was all there.' For both Sulston and Waterston, the justification of freely releasing information is not just that it's right, but that it works. 'It was not a theoretical concept, it was a pragmatic way of moving forward. The idea of free release came very directly from the need to get correlations between the physical and the genomic map, not from a theoretical idea,' says Sulston.

Waterston gives another reason for the release of genome data, which emerges from the nature of genes, and from the way they function in a cell. One gene hardly ever does only one thing, any more than one word in a language has only one meaning. Even when a gene produces only one protein, the protein may be involved in numerous different networks of cause and effect within the animal, which makes it of interest to all sorts of different investigations. So there are lots of reasons why different teams might have chased the same gene. 'And when you had to get the gene for yourself, then they'd be fighting to get the gene. They'd both

come at it from different directions and they ended up
trying to sequence the same gene. But as soon as you gave
them the gene, more often than not they ended up pursuing
different directions on it. It was just the bottleneck, you
know; once they got that, then they could pursue their own
dreams in their own way.'

By 1991, it was obvious that the only real obstacle to
sequencing the whole of the worm genome would be
money. The project was hugely expensive, and would never
be directly profitable. The money they had got from Watson
and the MRC would fund only the first 3mb of the sequencing. They would need to do thirty times as much if they
were to complete the worm, and it was by no means obvious where they would get it, since the first money that had
been given them was part of the human genome project
and it had already been enough to establish that large-scale,
speedy sequencing was feasible.

At this point, Lee Hood, whose lab at Caltech had been
the Alma Mater of ABI, the sequencer firm, put Waterston
and Sulston in touch with some serious money. An
American who had made a fortune in the shoe business,
Rick Bourke, was determined to do it again in biotechnology. What he seems to have planned was a firm that would
become the Microsoft of the human genome: by employing
the very smartest people around, and moving them all to
Seattle, he would end up with a lock on the most valuable
data in the world. The link with Microsoft went deep. Lee
Hood had recently moved to the University of Washington
at Seattle, to head a Microsoft-funded biotechnology
department; and the Bourke money was supposed to fund
the department's commercial arm.

Maynard Olson, who had invented the YACS, moved to
Seattle to the new department. So it did seem that they
would not be losing academically by moving there. A lot of

money was on offer: salaries and stock options quite outside Sulston's experience, and Bourke seemed to be offering the chance to get on with the worm as a project alongside the human sequence. But the longer they talked to him, the more clearly they realised that he did not care about the worm at all. 'We would simply be the scientific directors of a commercial sequencing operation.'* They turned him down.

It is an extraordinary illustration of the purity, and the single-mindedness of the project. There was nothing about worms that innoculated people against greed, yet Sulston, Coulson and Waterston were not seriously tempted by Rick Bourke. They wanted to do their own project, and they wanted money for that, not for their own sakes. Bourke was furious. 'John, I do hope this isn't going to do you any damage,' he told Sulston when the negotiations broke down. But it was Jim Watson who lost his job as a result.

Watson had been extremely worried by Bourke's attempts to sign up the world's best and most experienced sequencers to a private firm. He believed in international science (and he had himself become rich without needing to grub for it). So far as he could see, Bourke was trying to destroy a publicly funded project for private gain; and he let Bourke know it. Bourke thought Watson a dangerous reactionary who was standing in the way of American business, and he made sure his opinion was known on Capitol Hill and especially by the newly appointed boss of the NIH, Bernadine Healy, who had already locked horns with Watson over the NIH's decision to apply for patents on portions of certain human genes discovered by Craig Venter. Watson, when told of this plan at a public meeting, denounced it at once as 'sheer lunacy', which did not endear him to the new boss.

* Sulston, quoted in *The Common Thread*, p. 6

All through the winter and spring of 1992 Watson lobbied ferociously to keep the genome project international and publicly funded. He urged anyone who asked him (and some who hadn't) not to work for Bourke, who complained to the NIH about this interference. Lee Hood contacted Healy to complain about Watson, too.

But, while Watson's enemies were gathering in Washington, he was persuading his allies in England and elsewhere to fund Sulston and Waterston enough to keep them happy. To some extent the whole thing was based on a misunderstanding. Sulston really would not have gone to work for anyone who was not interested in allowing him to finish the worm. Watson had scared himself into thinking Sulston and Waterston would go to Rick Bourke, just as he had scared them into thinking, three years earlier, that he might fund the fly project instead of the worm. But Watson acted with extraordinary vigour to still his own fears. On a flying visit to London he managed to persuade the Wellcome Trust to start a human genome sequencing centre – something which would leave money over for Sulston to finish the worm. The money involved dwarfed the earlier applications: Sulston reckoned they would need £50 million and a purpose-built centre to do the work in. But the Wellcome Trust had just seen its income shoot up, because most of its assets were in shares of Glaxo SmithKline (GSK), a giant pharmaceutical company which had ingested Glaxo Wellcome, and GSK had invited the first patented, effective Aids drug. So the director, Bridget Ogilvie, found herself needing to spend £200 million a year just as Jim Watson was looking for people to fund an international sequencing effort.

Sulston jokes today that the Sanger Centre was built to sequence the worm. It wasn't. It was built to sequence the human. But I really don't know whether Sulston and

Coulson would have attempted the human genome had they been able to do the worm without it. This wasn't exactly humility: they thought they could do the human but this confidence did not affect their determination to see through the job they had started; and a belief that thoroughness was an essential part of good science. The Wellcome Trust would not fund the worm work, and the MRC could not afford the human work. But between them, they came up with £60 million to build a new sequencing centre in parkland south of Cambridge and to finish the worm there. The plans were completed in summer 1992; and everyone moved in over the next year.

While this was going on in England, Bourke had his revenge in Washington. He wrote to Healy complaining that Jim Watson was sabotaging American businesses. In response, Healy mounted a formal investigation into Watson's possible conflicts of interest. Like almost every first-rank molecular biologist except John Sulston, Watson had shares in a number of biotech companies. He sat on the scientific boards of two companies and a research institute, and he held stock in at least nine biotech or drug companies (among them SmithKline Beecham). Obviously, he would therefore benefit from the growth of the biotech industry in general, but there was no conflict of interest with his job as the head of the genome project. None the less, Healy pointedly refused to issue a statement clearing him of suspicion of impropriety, and Watson resigned and returned to Cold Spring Harbor. The real conflict had, of course, been over gene patenting and the question of whether the American government should adjust its scientific policies to benefit American companies or whether it should continue to cooperate in public projects. That war would continue throughout the lifetime of the human genome project. But Watson, though he paid with his job, won his battle to keep the public genome project going with the best people in the field; and they, the worm group, won theirs too.

The Sanger Centre was built in 1993 in the grounds of Hinxton Hall, just on the borders of Cambridge and Essex. It is sturdy and shiny, made with lots of plate glass and the kinds of glossy white material that could either be metal or a very modern plastic. You have the feeling that the only natural materials in the place are safe in petri dishes or scrupulously labelled freezers.

Though it is only about ten miles from the old MRC building, it belongs to a new century and a new kind of biology. It looks like a computing centre or a modern factory because that is really what it is. The space, the cooled air and the money, are all needed for the machines, rather than their attendants. The Sanger Centre was built less than thirty years after Sydney Brenner first decided that the worm would open the whole of biology to anyone with two toothpicks and a petri dish, but the last few reaches of the sequence needed resources unimaginable when he started. In 2002 the Centre spent about £10 million a year on computing resources – and of course the smallest and oldest computer in the place is incomparably more powerful than the Modular 1 that Brenner and White had started with. Even the pagers people carry around probably have more memory than that had.

The new sequencing factories grew faster and faster. The first three years of the worm sequence were taken up doing the first 3mb of genome; but the next 95mb took only another six years. Everything else speeded up at the same time as the worm did, but the most important changes came in the speed of human sequencing.

Sulston and Waterston found themselves locked in a really serious race with Craig Venter, the brain researcher who had urged on them the merits of ABI's sequencing machines. Along with Mike Hunkapiller, one of the founders of ABI, Venter set up a private company, Celera, to sequence the interesting bits of the human genome. Everything about Celera was repugnant to Sulston. He did not trust them scientifically or morally. The ambition to own and patent the human genome seemed to him futile and wicked. He thought the Celera group were offending against scientific virtues by hyping what they had done and what they expected to do. He himself had built his reputation within the LMB by saying little, but ensuring that what he said was absolutely true. That is the polar opposite of the technique required to raise a few hundred million dollars on the stock market. Celera's idea that they could tell in advance which were the interesting and profitable genes, and leave the publicly funded effort to toil in their wake, doing the dull and repetitive stuff, was not merely a sting to Sulston's pride but an offence against his idea of science. His argument would be that you cannot properly work out what is on a genome until you have all of it; and that to assume 'junk' DNA has no function because we can't see what the function might be is simply arrogant.

In all this, curiously, Brenner was more on Venter's side. Though he had encouraged the mapping of the worm, and helped to get money for sequencing it at a time when the Thatcher terror was emptying university biology departments; and though he threw himself into studying the genome of the puffer fish, he is disdainful of the Human Genome Project. 'I'm not very excited by that. I don't think it's science, actually.'

Brenner's disdain was well known, and not very well understood. Richard Durbin, who sees himself as a distant protégé of Brenner's, says, 'The sense I have is that, as he's got older, his time horizon has shrunk. I think he wants to achieve things in his working lifetime. I don't know if this is totally fair, but I think he thought it would take longer than his time allowed for to do this whole thing properly. When he turned out to be wrong, he got a bit grumpy about it.'

I think, though, that Brenner's argument is really that the collection of facts has far outstripped the capacity of theories to distinguish the interesting and important ones. He feels that if the genome sequencers had been subjected to the weekly interrogations of an LMB meeting – Why are you doing this? What do you hope to discover? – they would not have been able to give satisfactory answers. It is quite possible to see the huge effort that was made to collect the whole sequence of the human genome with the same kind of bewildered exasperation as comes up when Brenner's old colleagues remember his affair with the Modular 1.

While Brenner doubted the ends of the project, the disputes with Venter centred on the means. Venter had first gone private after failing to get a grant from Jim Watson to sequence human chromosomes by what looked like the quickest and easiest way, only looking for genes the cell had already found for you. Instead of using computers to sift through the millions upon millions of sequenced base pairs, trying to find coding patterns, Venter's technique (invented at MIT in the early Eighties) was to find messenger RNAS first. These are the little strings of RNA that carry the sequence information from the genes, where it has been read from the DNA, to the ribosomes, where the proteins will actually be assembled. If you have a fragment of messenger RNA, you know it will code for a protein. You also know that it will have been topped and tailed, and had all the introns trimmed out, so there is nothing there that does not code for genes. The trick is to collect these fragments of RNA, and use them as a template to make a copy of the original DNA.

Only at this point need you sequence the newly reconstructed gene; and even then you need only do three or four hundred base pairs at each end. Armed with these fragments, ESTS (expressed sequence tags) you can look for the rest of the gene in public databases, where it may well have been lurking, unrecognised. It is a wonderful technique, but if it is to be efficient it does depend on the existence of a sequence database, got by the unglamorous Sanger Centre methods, which can be used as a sea in which to trawl for the gene fragments you have recognised. It will not find the sites upstream on a gene which are used to switch it on or off. Nor will it even find all the genes in an organism, because mRNAs are produced only when genes are expressed – switched on by the cell – and many genes are expressed too seldom or in too limited a range of cells to be caught that way. It is no coincidence that Venter first found the method useful when working on brain tissue, where gene expression is most vigorous and varied, at least in organisms that have a brain.

Venter pitched his technique as a method of getting all the interesting genes without wasting time on the 'junk'. It is in the nature of the random sequencing methods the Sanger Centre used that the genes can come like ketchup from a bottle: very very slowly, and then all in a rush. The cDNA technique provided early successes. In 1990, Venter did some work on the worm on his own, using these cDNA techniques. Sulston, when he heard of it, saw it as a piece of simple imperialism. Venter seemed to be demonstrating that he had a technique which would work better than Sulston's, even on Sulston's preferred organism. So Sulston and Waterston between them got some worm cDNAs from Marty Chalfie's lab and in 1992, when Venter published his work on worm ESTS in *Nature Genetics*, they had their own paper ready to go in back to back with his.

After that, they returned to steady sequencing. They had made the point that they could use Venter's preferred technique just as well as he. That didn't mean they thought it was the best. They did, however, encourage Yuji Kohara, who had spent a year in Cambridge in 1986, to work on cDNAS as a way of identifying genes in the stretches already sequenced: it was another huge labour but in the end he found nearly 7,500 genes by looking at nearly 70,000 ESTS.*

For the next six years, things got faster and smoother. There were steady small improvements in the machines they used: each year more raw DNA could be sequenced more quickly; at the end of 1998, they decided it was done. The decision was to some extent arbitrary, though less arbitrary than the corresponding choice for the human genome. The overwhelming majority of the genes they thought they had found had never been identified in nature, but only spotted as likely by the computer programs that analyse the sequence data. The generally accepted figure is something around 19,000 genes. More detailed sampling suggests that at least 17,000 exist, and few people would regard 20,000 as an unreasonable guess, taking in all the minor and hard-to-find genes that code for small and specialised bits of the cell's machinery, like RNA. But these are still guesses, though well-informed, and as late as 2001, Ian Hope, a well-respected worm scientist from Leeds, was arguing that the consensus overestimated the number of genes by as much as 20 per cent.

The sequence is not only ambiguous. It is also not quite complete. There are still a few tiny bits being worked on; but they were confident by the end of 1998 that these didn't matter; besides, at some stage in a project like this

---

* Actually, 67,815 ESTS had produced 7,432 genes by the time the sequence was published (*Science*, 282).

there has to be some kind of ceremony to reassure the world that all that money has not been wasted.

The completed worm sequence was announced on 11 December 1998. The MRC had spent £16 million on the sequence alone; it had cost the American funding bodies about as much again. The event was celebrated with a special issue of *Science*, and a press conference where Bob Horvitz compared the project to the moon landings. That looks like absurd hype, and the claim in the *Science* paper was less flamboyant, though no more modest when closely examined. 'As a resource, the sequence will be used indefinitely, not only by *C. elegans* biologists but also by other researchers for comparison with and the interpretation of other genomes, including the human.'

The sequence will be used *indefinitely*. Given that it is the fate of almost all scientific research to be forgotten as soon as it can be improved on, this is a claim to immortality on a Shakespearean scale: 'So long as men can breathe or eyes can see, so long lives this.'*

'And this gives life to thee', continues Sonnet 18. But does the sequence give life to the worm? The great paradox of the sequencing effort is that it is just as true that the worm gives life to the sequence as that the genes give life to the worm. 'DNA isn't life: DNA in a fertilised egg is life,' says Sulston. Computing an organism from its genes is still something which only a living organism can do.

For the next two years, as the hype surrounding the human genome rose to fever pitch, such distinctions could not sensibly be made. Sulston and Waterston continued to drive their human genome project forward in competition with Craig Venter's Celera. The Celera share price hit

* In a future world shaped by molecular biology, 'So long as eyes can see' may be a larger bet than 'so long as men can breathe'.

$242.00 in February 2000 (it was around $10.00 in August 2002). The race to the human genome sequence was officially declared a dead heat in summer 2000; and the draft sequences were published in February 2001. Work is still going on to finish the draft, which was announced at a state much less complete than the worm's had been when it was announced. The human clone libraries, which allow people to get at the physical DNA the sequence displays on their computer screens, are in terrible disarray compared to the worm's. Donna Albertson, who now does human research, says that about 15 per cent of the samples referred to in the physical map have been mislabelled or thrown away, because all anyone cared about in the Nineties was getting the sequence data out, and once it had been transferred to the databases they no longer cared about the physical substrate. Yet it is only the links with the physical map that make a sequence worthwhile.

Originally, this book was to end with the sequence. It obviously was not the end of the worm, but it was one culmination of Brenner's original labours. In thirty years, they had discovered every single cell in the worm and its history; every single nerve cell; and now they were close to every single gene. All this had been done at a level of detail which is even now hard to comprehend and must have been ridiculously optimistic in 1965. What more could there be to discover about an animal so small and so limited in its behaviour?

10

The End

The first worm meeting ever held, at Woods Hole, had twenty-four people present; most of them were only part-time worm biologists anyway. Bob Edgar got drunk and made a speech about how the community spirit was disappearing. The 2001 worm meeting had 1600 people gathered on the hillside campus of UCLA. The private campus police patrolled to ensure no one drank alcohol except on one approved evening. Everyone I talked to told me soberly the community spirit was disappearing. They were all having a really good time.

They were young and tightly focused. I caught a loving couple in front of me swapping notes during one particularly technical lecture, and then noticed that their notes were chemical diagrams. There was a nervous, hungry confidence in the air which did not seem like the descriptions I had been given of the MRC; it was missing the element of fun. That was understandable. These young scientists were in a very different position from the young men who, thirty years ago, had set out to understand the worm. Their field was far more crowded. They were far more rigorously examined and managed than their predecessors; and the demand for paper qualifications is higher in modern science, as in everything else. Alan Coulson had gone to work as Fred Sanger's assistant with nothing better than an HND degree from a polytechnic. It is difficult to imagine someone with that background getting his first job working for a Nobel prize-winner today. Lower down the ladder, the technicians Muriel Wigby and Eileen Southgate had started work as school-leavers, with no tertiary education at all. But when they retired they were both replaced by PhDs doing pretty much the same job.

Any scientist has to pick a problem to work on which is going to be large and hard enough to be interesting, but not so large and so hard that it remains insoluble; and this choice is in the nature of things a gamble. It is part of the difficulty of scientific problems that you don't know how difficult they are until they have been solved. But the margin for error keeps shrinking. Richard Durbin says, 'When you're young now you can't afford to get [the choice of problem] wrong badly, which I think is bad for science. Certainly, the LMB's approach was to take people whom they regarded as the right sort of people – bright – and to allow people who were good to do risky things and fail and have another chance because it is quite risky, science.' Once you have the right problem, you have to solve it before anyone else. That is sometimes a problem even for people who are 'bright' in the sense that the word is used in Cambridge, meaning someone who has no problem getting any degree he really wants.

As the stakes have risen, so has the power of the tools available. Durbin himself was partly responsible for one of the most powerful of them. He has spent his whole working life shuffling between computers and biology: he started as an undergraduate mathematician, and then did a PhD with

John White on the worm's nervous system. *The Mind of a Worm* listed all the 5,000 connections they found between nerve cells. But it could tell nothing about the traffic among them, or why nerves formed the connections that they did, since often they will touch, or run alongside each other, without exchanging signals. *C. elegans* had been chosen in the first place because it was small enough for slices to fit under an electron microscope, but this fact, which made it possible to trace the nervous system's physical existence, was also what prevented them from tracing its activity. The nerves were too small to attach any kind of monitoring apparatus. Durbin hoped to establish mathematical rules setting out when and where connections might be formed. The effort wasn't completely successful, but it led him to map the nervous systems of worm embryos just as White had mapped the adults.

From that work he moved sideways into helping John Sulston write the software that sorted out the jigsaw of fingerprinted cosmids; then, with a French physicist-turned-biologist, Jean-Thierry Mieg, he wrote AceDB, the software that holds all the information about the *C. elegans* genome sequence. AceDB, and its successor, Wormbase, are really where all the knowledge about the worm is now to be found. They mark the point at which scientific knowledge stopped being something which could fit into textbooks and had to migrate onto the web.

Thirty years ago, nothing was known about *C. elegans* except for the inaccurate stuff in the textbooks Brenner read in Cambridge; in 1986 everything known could be fitted into one volume, and it could just about fit there for another decade. But by 2002, *C. elegans II*, the final textbook from 1996, had been republished on the web, where it was very much easier to read and navigate, if harder to annotate, than in the 1,200-page paperback edition published in 1996. Out on the web, too, you could find the complete text of *The Mind of a Worm*, and of Richard Durbin's PhD thesis which had carried that work forward. But the flood of new work rose inexorably. The 2002 worm meeting had more than a thousand new papers read to it. The abstracts of all of them were published on the web.

AceDB had not been published on the web, though the software and the information was distributed on the Internet in fresh editions every six months or so. Wormbase is accessible for any computer, and here, in the densely layered annotations, you can see what genomics really means. None of the layers on its own would mean very much, and all could perhaps be fitted into separate books. But the combination of all these different views of a gene, none more than a couple of mouse clicks away from all the others, really does show a new way of looking at the world. The raw sequence that made it all possible is almost the hardest layer to find, and certainly the most uninformative.

The heart of genomics is the display you come to when you first type in the name of a particular gene. Here there is a series of parallel lines, showing where the gene is in a chromosome, where it is in a library of clones, how it is broken up by introns, and how it can be spliced where this known. There are also displays to show which probes can be used to find it, what happens to worms in which it has been knocked out, which proteins it makes, what scientific papers have been written about it, and how closely related it is to genes in other animals, including humans. Obviously, software like this will not answer all your questions unless they are very well informed ones. But to wander around it, and consider the huge anonymous labour that went into it, is to come to something very close to a modern cathedral.

All the early pioneers' work has now been reconstructed in this virtual reality. The painstaking work on embryonic lineage that Sulston did is now automated and visible in a different program, which lets you watch the development of an embryo in three dimensions. Each cell can be seen swelling and splitting in real time, or growing dull and dying if that is its fate, while its genealogy and future descendants are displayed alongside the video window.

The pictures of the embryo cells have the grey appearance of all Nomarski microscopy, like porridge under glass. But another technique has been developed to allow worms to star in technicolor movies. It is mostly the work of Marty Chalfie, who found a way of splicing a gene from Pacific jellyfish into worm DNA. The GFP gene – the name stands, prosaically, for 'green fluorescent protein' – makes a protein which glows under ultraviolet light. The gene for this was found in the early Nineties by a man named Doug Prasher. His is a cautionary tale about the importance of mapping and sequencing, for it took him three years to find the gene in a library of cloned jellyfish DNA, and by the time he had done so, his funding was exhausted. So he passed the gene to Chalfie, at Columbia, who was able to take Prasher's sequence and check that it passed the really critical test and made the fluorescent protein even outside the jellyfish.

Chalfie discovered how to get GFP working, first in *E. coli* and then in the worm. Now it is used in any animal you can imagine. GFP glows luminescent green under ultraviolet light but doesn't do anything else, which makes it a harmless but highly efficient way to watch what is happening inside a living worm. If you want to know where and when a worm uses a gene, you splice the GFP marker sequence next to the sequence of the gene that you are tracking, so that both proteins are switched on and off together. The worm can then be filmed under UV light and all the cells where the gene is being expressed, and making its proper protein, show up as if marked with a green highlighting pen. This method can even be used within cells, to watch the ways chromosomes are pulled apart when they divide.

Leon Avery, who publishes the *Worm Breeders' Gazette* on the web, says that the pace of change in biology is nearly as fast as it has been in computing. 'Every year, or every two years, we throw out the tools we were using two years before for ones that are ten times as good.'

Ten years ago, he says, when you wanted to find where a gene was expressed in a worm, 'you cloned the gene, which took a while there, and then you expressed the protein in *E. coli*, you injected into bunnies, made some antibodies and then you stained the worms to the antibodies. That's how you got expression patterns. So, beginning to end, if everything went really well, that was a few months.'

Now one of his students has worked out a method which starts from the sequence in Wormbase and within twenty-seven hours the new GFP-flagged gene has been constructed and injected into worms. Within a week – a couple of generations – there will be thousands of worms in which the gene you are interested in can be studied. Once this has been done for the first time, some of the worms are frozen in suspended animation so that the mutant strain is available to anyone who needs it.

It was not simple to get the GFP-tagged genes into a worm. Judith Kimble tried for years in the Eighties to inject modified DNA into the gonad of the worm so that some of it would be taken up by the maturing eggs and be incorporated into the next generation of worms and their progeny. The problem was eventually solved by Andy Fire, a biologist in Maryland: the technique is not to stick a needle in the worm, but to lower the worm onto a glass needle until the cells surrounding the gonad can be filled with new DNA.

At first the technique was used to do the obvious trick of injecting fresh DNA into the worms. Then it was modified to try to knock genes out: in theory, you could target the single strands of messenger RNA that serve as the final, edited template for a protein by making exactly complementary strands, which would match them all along their lengths, scrunchle up into a double helix, and become impossible for the cell to transcribe. In practice, this technique never worked as it should have done, and no one quite understands why. Andy Fire, again, found the way round, and came up with a trick which allows targeted knock-outs of any gene which can be sequenced.

Like Fred Sanger's dideoxy sequencing, Fire's RNAi depends on the subversion of natural defence mechanisms against viruses. Sequencing uses the restriction enzymes that chop up DNA viruses when these appear in the cell. But not all viruses are made of DNA. Some – HIV is a notorious example – are transmitted as RNA and the cell has different and specialised defences against them. RNA viruses normally appear in cells doubled back on themselves, like hairpins. If these doubled, complementary strands of RNA are detected, the cell has enzymes which use them as templates to destroy any matching single strands of messenger RNA. This would normally be a very efficient and precise way to prevent the viruses from using messenger RNA to subvert the cell's protein-making ribosomes to make rival proteins. But if you can mimic the sequence of one of the cell's own genes in double-stranded RNA, and inject that into the animal, the messages from that gene will never make it to the ribosomes. They will be chopped up and destroyed while they are still in messenger RNA. This provides an enormously subtle and powerful method of knocking out individual genes.

RNAi does not work only in animals which have had their gonads injected. The strangest and most profound thing about it is that bacteria which have taken up the RNAi can be fed to worms, and they will absorb it, and stop production of whatever protein the RNAi has specified.*

Taken together, these techniques allowed traditional genetics to be turned inside out. Instead of working inwards to the genes from their bodily effects, it was possible with the new 'reverse genetics' to start with a gene and work outwards to its effects on the body. The transformation of the worm into genetic Meccano was complete. Almost everything that goes on inside a worm can be monitored, changed and debugged. It has become the most completely understood animal in history.

Paradoxically, perhaps, this has made it less interesting. The most profound discovery hidden in Wormbase is reached through the buttons that let you compare worm sequences with those of other animals. It turns out that many of the fundamental processes of life in *C. elegans* are interesting because they also operate in *Drosophila* and even *Homo sapiens*; since they are of greater interest to us when they work in *Homo sapiens*, quite a lot of worm genes were first discovered in humans.

How like a human is a worm? A great many of the current worm researchers are working at human cancer research institutes; at a genetic level, there is clearly a great deal of similarity between a human and a worm; and, for that matter, between a human and a banana, with which we share half our genes. But what does this mean in practice? How can a creature with no brain, no bones and no immune system possibly have anything practical to tell us about ourselves?

Human beings and other vertebrates are not descended from nematodes. We are very distant cousins, probably

---

* RNAi works on humans, too. In an age of bacteriological warfare, that's the most frightening footnote I've ever written.

more closely related to lobsters or to spiders than to *C. elegans*. But even such distant cousins must have a common ancestor; and that common ancestor, whatever it was, solved most of the problems of becoming multi-cellular seven or eight hundred million years ago, and we, all its descendants, have inherited those first solutions. That is why it was worth spending thirty years looking at twenty-two cells in a worm's vulva. The mechanisms Bob Horvitz was investigating when they go wrong in a bag of worms turn out to be the same in all animals we know of, including ourselves and flies. The genes that he discovered are shared by worms and humans won him his share of the Nobel prize in 2002. They are not the only mechanisms in a human: more complex animals have more complex internal signalling networks. There is a lot more going on inside a mouse than inside a worm, and the increasing complexity of our genome reflects the increasing sophistication of the messages that must be carried between and within cells, or of the proteins that carry them. But the mechanisms that are found in a worm will also be found in a mouse, and some of the proteins that develop a worm's vulva are also used to develop a human brain.

The way this happens goes right back to the first studies ever made of the worm, and the ways in which the nervous system grows. The ways in which the nerves of a worm extend through the body and form their connections are not entirely the same as those used by other animals. For one thing, in a worm it is the muscle which grows towards the nerves that will control it, whereas in almost all other creatures with nervous systems this happens the other way round. But the nerve cells, as they grow, still have to find their way to where the muscles can reach them, and they still have to find their way to each other. Human and worm nerves both grow by following a cell which threads its way through the body to its final destination.

Most of these fundamental mechanisms have been discovered in humans by cancer researchers. Cancer is what happens when the cells of a body start to proliferate anarchically, without taking account of the signals that normally regulate development. If you find and clone a gene important in cancer regulation in humans, and put it into a worm, it will often do the same job there, which means it turns out to specify a protein which fits together with worm proteins as it should. For obvious reasons, the experiment is not done the other way round. But it would work.

The genes and proteins that have been conserved for hundreds of millions of years must be of central importance. When a protein stays almost the same through uncountable generations of creatures, this is not because it is uniquely durable; it must be liable to mutate at the same rate at any other. But hardly any of these mutations will have been copied, because the precise form of the resulting protein and the ways it fits into others are so important to the proper functioning of the organism.

Worms even show learning and boredom of a sort. Obviously these behaviours aren't the same as they are in animals with brains. But consciousness and intelligence come in layers, like everything else built by evolution. In worms, you can see the beginnings of complex behaviours which arise from the activity of very simple nervous systems. As an example of memory in worms, consider that, if offered a choice of temperatures, worms will try to move towards the one at which they were raised. Worms which grew at high temperatures will prefer them all their lives. The only exception comes when they have been starved, when they will lose this preference, and seek out different temperature zones. Both behaviours make sense in their native environment:

they are ways of tracking likely concentrations of bacteria, which also have their preferred temperatures. They work because abundant food is a stimulus to egg-laying, so a worm should have been born at a temperature agreeable to the local bacteria. On the other hand, if food runs out it makes sense to look for a different colony of bacteria, which may well thrive in different conditions.

They can even get bored of specific sensations. Normally, worms are attracted by some smells and repelled by others. You can measure how attracted they are by seeing how far and how fast they move towards chemicals. But if they are exposed for long enough to even the most attractive chemical, they will gradually stop being attracted by it. You cannot measure this change by watching how nerves grow, because they don't change when it happens. It is known exactly which nerve senses which chemicals; and even after a worm has grown bored of a particular smell, it will continue to react strongly to others which are detected by the same nerve.

Why this happens is still unknown, after thirty years of intensive study. Neither are the mechanisms known by which the worm can sometimes defer its aging, though this is the field of research most likely to propel the creature into the headlines. But no one doubts that these questions will be answered. The end is in sight when everything interesting in the worm can be understood, so that it becomes apparent exactly how the function of any protein within the worm arises from its shape and other physical properties and the ways they interact with the known shapes and properties of other molecules. The paradigm for such an understanding is the way in which the function of DNA as a carrier of hereditary information follows naturally from the way that its two strands match up together to make the double helix.

Reaching the same understanding with the signalling proteins that govern the ways in which worms – and humans – grow is hugely complicated; but no one I talked to doubted it would happen. Leon Avery said, 'We would like to be able to describe in molecular terms how every cell type is developed and ends up in the right place and I think, actually, it's not going to be very long at all before we have that.'

But Avery's ambitions went further. He wanted to know how the worm sustains itself, not just how it grows. This turns out to be a more complicated question. The end of the worm project seems to recede further the more we know: Brenner's ultimate plan could be expressed as a computer model in which the worm would be so accurately represented and so well understood that models of drug molecules could be tested on the model worm and come up with the same results as a test in the real world. That is at least twenty years off, according to most of the people I talked to; and it may not happen at all.

One theory is that the worm will never be understood because it will cease to be interesting. Phil Anderson, for one, believes that it will simply fade away, much as the phage did, because it will cease to be the best organism – though biologists say 'system' as if they were working with computers – for studying the interesting questions. 'Phage biology has kind of disappeared as a field, because people learned how to do the experiments, how to do approaches, and went off and applied them to other systems.

'Unlike my own lab, which is completely dedicated to studying *C. elegans*, a typical lab ten or fifteen years from now may have a portion of its work dedicated to *C. elegans* which interfaces with another portion dedicated to studying human genes and diseases and things like that. Phage biology did the same thing. People that worked on phage started studying *E. coli*. Then they started studying, just like

Sydney, *C. elegans*; and Benzer did *Drosophila*. But it's only now you can look back and say, "Well, you know, there's nobody working on phage any more."'

This idea that the worm will fade away is a common one. Richard Durbin reckons the worm has another ten or fifteen years at the forefront of biological interest before all the interesting and little-understood questions concern the processes that are only found in creatures more complicated. Underlying this is the belief that all biology will become one science, in which the organisms are used as facets of a greater whole, rather than being studied for their own sakes. It is in that sense that John Sulston says, 'When we understand the worm, we will understand life.'

Yet the worm retains its individual mysteries. What does it want with nearly 20,000 genes? That was the question raised by the last speaker at the 2001 worm conference, Jonathan Hodgkin. His answer suggests that the last thirty years of worm research have missed some vital points.

The worm's 19,000 genes are two-thirds as many as have been predicted for a human being, and much more than is needed for *Drosophila*, the fruit fly Brenner had originally dismissed as having a nervous system too complicated to study profitably.

Some of the reasons Hodgkin came up with are technical, but important, too: in bacteria, genes can be read off the sequence as a single string, which runs continuously from start to finish. But all other organisms have their gene sequences interrupted by 'introns', sequences which don't code for proteins and are edited out by the cell's machinery before the final stretch of messenger RNA is prepared that contains the exact template for a protein. This complicates life because it overthrew what had been one of the most productive ideas of molecular biology: the simple doctrine that one gene makes one protein. It turns out that one gene can make numerous proteins, depending on how the cell decides to edit it. There are reckoned to be an average of three different proteins coded for by every human genes; one gene (DSCAM) in *Drosophila* has been found to code for fifty different proteins and the theoretical maximum is much higher.*

But nematodes do far less of this 'alternative splicing' than most other animals. That may be one reason, Hodgkin suggested, why they need so many distinct genes. Even so, there is a huge number of nematode genes which seem to have no function at all. There are about 7,000 genes which have an identified phenotypic effect. That is only a little more than a third of all the genes people believe are there. So the question of what the remaining two-thirds are needed for becomes more urgent. It may not be possible to answer it, even in the most sophisticated labs imaginable.

In the laboratory *C. elegans* lives in conditions as carefully controlled as human ingenuity can make them. It has no predators – unless you count humans – and no diseases. It is fed abundantly, but only on one bacterium. Almost all the hazards of life in the wild have been eliminated. So when in 1986 some of the worms in Hodgkin's lab developed swollen, club-like tails, it looked like a new mutation. It took some time to realise that they were just constipated. They had been parasitised by a hitherto unknown bacterium, *Microbacterium nematophilum*, which lived in their rectums and multiplied into a solid plug.

As an energetic geneticist should, Hodgkin started to look for mutants resistant to this infection, and he found at least twenty genes which in some ways affected the worm's ability to keep the bacteria out. It is worth noting that the

---

* A thousand, according to Hodgkin.

sanctuary *M. nematophilum* chose has considerable attrac-
tions to a bacterium. Not only does it assure a ready supply
of food, but it is also one place where it is safe from being
eaten itself by *C. elegans.*

Forming these mats or plugs seems to be quite a common
defence mechanism against nematodes among bacteria. In
at least one instance, it has disastrous consequences for
humans. *Yersinia pestis*, the bacterium which causes the
Black Death plague in humans, has been shown to form
mats or plugs in the presence of *C. elegans*. Unlike *M.
nematophilum*, it does so outside the worm. The effect is
much more dramatic. The mats of plague bacteria form a
rubbery skin over the worm's mouth and stop it eating, so
very few larvae exposed will survive to adulthood. The
same mechanism explains, indirectly, why we get the
plague. The bacterium forms similar plugs inside the stomach of an infected flea, so that it spits infected blood into
the humans it bites. So we catch the Black Death because of
a defence mechanism evolved to protect bacteria from
nematodes like *C. elegans*. Perhaps the worm has had its
revenge after all.

Hodgkin, however, has a serious point to make. If the
struggle between the worm and one particular form of bacterium can affect twenty hitherto unknown genes – and
nothing is known about most of them except that they
exist – the complexities of the worm's real life in the wild
will involve hundreds, possibly thousands, of genes to help
it deal with the enormous range of bacteria it tries to eat,
and which will attempt in turn to parasitise or otherwise
infect it.

Because of this, he believes, we will never reach a full
understanding of the worm. In fact, there is only one
species, he says, we can hope that biology will completely
explain, and that is the most complicated one on the planet,

Homo sapiens. 'Only we are in a position to report on every disease or toxin we encounter. Similarly, only we can adequately monitor our own physiological and genomic responses to Arctic blizzards, tropical heat, psychological stresses and social pleasures.'

Most of the species on earth may well go extinct in the next century. The worm will survive almost anything we can do to ourselves and the planet: so long as plants rot down into compost and bacteria eat them while they do so, there will be nematodes to eat the bacteria.

After I wrote that, I realised I had never seen *C. elegans* in the wild. I had fallen in love with a pattern of writhing blue shapes on a wall in Jesus College, Cambridge, where someone giving a talk on the future of science had placed a petri dish on an overhead projector, and let us watch the worms as they moved around on screen while he talked about the ways in which these creatures would soon be entirely understood. I had watched innumerable films of worms since then, and peered at them down microscopes in the Sanger Centre. But they were all worms which looked like mechanisms. Many of them had been highlighted with GFP; some had laser pointers indicating their genitals. Even their eggs, in this perspective, were round like shrouded cogwheels. None of them looked as they might in nature.

Curiously, in all the worm literature, there was very little instruction on how to go about catching one in a garden, or a flowerpot. If a scientist wants a worm, the *C. elegans* Genome Centre will post any one of 4,000 different identified mutations, or even a certified wild-type worm. Why look in the garden for something unclassifiable? But eventually, I found some instructions from Phil Anderson in the *Worm Breeders' Gazette*, and a slightly more elaborate set in an old *Scientific American*. They needed a sieve, some cheesecloth and a container.

By this time I knew so much about the animal, and it seemed to me to encompass so much knowledge and so many things, that it was astonishing to realise how insignificant it looks in the wild. To catch real and abundant nematodes, you need to dig down into the soil for a couple of trowels' worth of earth, and put this through a coarse sieve, a process which may occasion domestic disagreement: the *Gazette* prefers pie-tins with a mesh base. In any case, the idea is to remove stones and break up lumps of earth. The sieved earth is then wrapped in slightly porous material – kitchen paper does fine – and placed in a container with a little water.

Over the next few days, all the nematodes in the earth will emerge through the paper filter into the water, a process when the transparency of *C. elegans* is a disadvantage. How did anyone ever discover the worm in the first place? On the third day, we found a few white threads which bent and slowly straightened when touched with a needle. With some difficulty, we fished them out and arranged them under a cheap microscope. They looked pallid and without the blue cast imparted by professional microscopes. They were a watery grey with dark lines marking their muscles. Their heads moved sluggishly from side to side; and I realised that I had no idea whether these were *elegans* or *briggsae* and no clear idea of how to check. Everything about them was mysterious, even their identity. After months of immersion in helminthological literature, in which I had never been more than a page turn or a mouse click away from the most refined and detailed knowledge of its molecules, I was once more confronted with the brute enormous strangeness of the creature. I felt as if I had been helicoptered back from the summit regions of a mountain, and was once more sitting in a teahouse at its base, with at last some understanding of the enormous distances and discouragements concealed within the picturesque view. From what I could see, Brenner had set out all those years ago with his toothpicks and his petri dishes to map and climb; and they had nearly done the whole damn thing.

I went to meet John Sulston in the pub that night. We talked about the end of the worm, and he said, as he had before, that his view was simple: 'When we understand the worm, we'll understand life.' That's not, I said, what most people mean by understanding life. It wasn't what he had meant, either. The understanding of life, in Sulston's sense (which is what has animated the whole of molecular biology) is not a philosophical or ethical achievement. It is not an answer to questions such as 'What should we do?' or 'Why are we here?' It is a technical question to which complete and indisputable answers must exist. Months earlier, talking in his office, he had put it slightly differently. 'DNA is not life. DNA in a fertilised egg is life. Huge difference. The question is whether we shall one day be able to make a synthetic cell which will take a synthetic piece of DNA. For if we can, then we shall have created life.'

'Of course,' he added in the pub, 'we're still a long way off.'

Note on Sources

For the people and some of the ideas and techniques involved, the canonical history of molecular biology is Horace Freeland Judson's *Eighth Day of Creation: Makers of the Revolution in Biology* (New York: Cold Harbor Springs Laboratory Press, 1996). I have relied on it for the prehistory. Sydney Brenner made a videotape in conversation with Lewis Wolpert, which was edited into a book, *A Life in Science* (London: BioMed Central, 2001).

John Sulston has a partial autobiography, written with Georgina Ferry, *The Common Thread* (London: Bantam, 2002). The best book on the early part of the Human Genome Project is Robert Cook-Deegan's *Gene Wars* (New York: W. W. Norton, 1996).

Other sources I found useful are Sydney Brenner, 'The Genetics of Behaviour', *British Medical Bulletin*, 29: 269–71, and 'The Genetics of *Caenorhabditis elegans*', *Genetics*, 77: 71–94; Riddle, Blumenthal, Mayer and Priess, *The Nematode Caenorhabditis elegans* (New York: Cold Harbor Springs Laboratory Press, 1988); Soraya de Chadarevian, *Designs for Life* (Cambridge University Press,

2002); and Rachel Ankeny 'The Natural History of C. *elegans* research', *Nature Review Genetics*, 2, June 2001.

On the worm itself, Soraya de Chadarevian, and Rachel Ankeny have both published on the web reprints of scholarly treatments of the early days of the project.

Almost everything you want to know about the worm itself is best learned online. All the classic texts have been placed there: John White's *Mind of the Worm* and *c. Elegans II* are both online in their entirety, and properly indexed, too. Wormbase, the complete database, contains complete cell lineages as well as the genetic and physical maps and the annotated sequence of the worm. All web references can be found at www.thewormbook.com/refs.

Index